OR SOMETHING WORSE

Nicholas Beuret is an activist scholar and has been engaged with questions of climate justice and ecological politics for thirty years. He is a lecturer in environmental politics and economic geography at the University of Essex. His work has been featured in Novara Media, the *Ecologist*, and openDemocracy. He lives in Britain, where he continues to write, teach and engage in environmental advocacy.

OR SOMETHING WORSE

Why We Need to Disrupt the Climate Transition

Nicholas Beuret

VERSO

London • New York

First published by Verso 2025
© Nicholas Beuret 2025

The manufacturer's authorised representative in the EU for product safety (GPSR) is LOGOS
EUROPE, 9 rue Nicolas Poussin, 17000, La Rochelle, France
contact@logoseurope.eu

The moral rights of the author have been asserted

1 3 5 7 9 10 8 6 4 2
Verso
UK: 6 Meard Street, London W1F 0EG
US: 207 East 32nd Street, New York, NY 10016
versobooks.com

Verso is the imprint of New Left Books

ISBN-13: 978-1-80429-985-2
ISBN-13: 978-1-80429-986-9 (US EBK)
ISBN-13: 978-1-80429-987-6 (UK EBK)

British Library Cataloguing in Publication Data
A catalogue record for this book is available from the British Library

Library of Congress Cataloging-in-Publication Data

Names: Beuret, Nicholas author
Title: Or something worse : why we need to disrupt the climate transition /
Nicholas Beuret.
Description: First edition paperback. | London ; New York : Verso Books,
2025. | Includes bibliographical references.
Identifiers: LCCN 2025013280 (print) | LCCN 2025013281 (ebook) | ISBN
9781804299852 paperback | ISBN 9781804299869 ebk
Subjects: LCSH: Global environmental change | Carbon dioxide mitigation |
Climatic changes – Economic aspects | Environmental protection – Social
aspects | Green movement – Economic aspects
Classification: LCC GE149 .B49 2025 (print) | LCC GE149 (ebook) | DDC
304.2/8 – dc23/eng/20250617
LC record available at https://lccn.loc.gov/2025013280
LC ebook record available at https://lccn.loc.gov/2025013281

Typeset in Garamond by Biblichor Ltd. Scotland
Printed and bound by CPI Group (UK) Ltd, Croydon, CR0 4YY

Contents

Acknowledgements

To acknowledge someone is not just to recognize their contribution, or to express gratitude. It is to confess to an alliance, a filiation. This book, including the excess of material that was cut or put aside, is the work of a series of alliances and collective, generative acts. It has come from arguments, conflicts and debates; conversations and the collective writing of plans and texts. It has been done with others, not calmly, but in anger against the state of things.

It could not have come to completion without my being able to call on the generous contributions of a number of people. My partner, Camille Barbagallo, has worked through all this material and more with me, from late night conversations and arguments to editing the manuscript despite having no time to do so. David Harvie generously commented on the entire draft and good naturedly put up with endless esoteric messages from me as I worked my way through chapters. Stuart Melvin and Kylie Benton-Connell offered insightful feedback on early chapters that helped shape the final argument. Both editors at Verso, Rosie Warren, who supported the project from the beginning, and

Tom Hazeldine, contributed much to the text for which I am deeply grateful.

Many of the arguments here owe a great deal to conversations with friends and comrades. Esther Lutz Davies, Bue Rübner Hansen, Samuel Barbagallo, Simon Pirani, Owen Espley, Paul Rekret, Angus O'Brien, Silvia Federici, George Caffentzis, Nick Cullen, Alessio Lunghi, Pascoe Sabido, Les Levidow, Leon Sealey Huggins and Emmanuel Blondel have all marked this book in crucial ways. I owe a great deal to all my comrades in the Institute for Commoning – Adam Barker, Emma Battell Lowman, David Harvie, Amber Huff, Cath Long, Gareth Brown, Nathan Oxley, Patrick Huff and Claire Marris. Some of the key arguments of the book were previously developed through articles for Novara Media and the *Ecologist*, and I am grateful for the support and encouragement of Clare Hymer and Brendan Montague.

Writing this book would not have been possible without the love and support of my children, Azadi and Bastien, both of whom have given me more warmth and encouragement than I had thought possible.

All too often books on climate politics end on this note, turning from the authors' children to their hopes for the future. But the future is what we are missing. This book is dedicated to our struggle in the present, so that we all might do more than just survive.

Introduction

The Debate Is Over –
on to the Green Transition

We are in the midst of a profound social, political and economic transition. Despite the rise of the far right and climate denialist demagogues, despite covert campaigns by oil companies to weaken regulations and ban protests, and despite the seemingly relentless rise of carbon emissions, the transition to a so-called greener global economy is underway.

The central paradox of not only climate politics but all politics today is that this transition is also one into a much hotter future. The green transition is moving away from fossil fuels and into climate catastrophe. Between the professional optimists on one side, endlessly celebrating each increase in solar panel installation rates, and those declaring the transition nothing more than a lie on the other, is the reality of the transition. It is both already happening and won't stop climate change at anything we could regard as a 'safe' level.

The core argument of this book is that we are in the middle of a war of transition, but only one side is fighting. What is needed is for us to understand the shape of the transition, to map its contours and contradictions, as a process that is already underway. We need to do so in order to organise not against it, but through it.

This transition is the terrain of contemporary class and social conflict, one that is reshaping our everyday lives and radically transforming political strategy and action. Business and government are remaking the social contract, squeezing our living standards and transforming work, while forcing us to pay for the transition. The transition to a green economy is returning us to a state of economic nationalism and neo-mercantile politics, underpinned by the shadow of 'security interventions' and a renewed round of resource colonialism and outright war.

The unsettling of previous industrial paradigms and governmental regimes, not to mention the collapse of past ideologies and hegemonies, presents us nevertheless with an opportunity. A moment of transition is one where old orthodoxies and contracts come undone, and where new ones are yet to cohere. It is a political terrain where any conflict can seize not only the imagination but the agenda, driving profound, previously unthinkable change. In a war of transition everything becomes immediately political, while social forces rapidly emerge in seemingly impossible situations.

The struggle around the climate crisis is measured in degrees. And while each degree matters, the stakes of the transition to a decarbonised economy have been profoundly underappreciated. Climate change doesn't occur with one catastrophic event but arrives instead as a series of crises, each provoking political reactions and attempts by capital to profit from disaster. As the climate degrades further, this risks becoming a downward spiral. Within this spiral are the changes most people experience in their day-to-day lives – in the workplace, their wallet, the supermarket, in the decline of local services and in what people think they can and can't expect from the future.

Both the Left and the climate movement are at an impasse. Hopes for the revival of a 'green' social democracy have faded. Electoral politics largely appears to have been lost to the reactionary forces of the far right. Much of what passes for climate action is little more than symbolic, obsessed with the spectacle of dissent and building public support for 'acting now'. Radical action itself seems impossible. Instead, climate action has reached the dead end of policy obsession. Or, perhaps worse, it has retreated into arcane questions of strategy far removed from the reality of either the transition or the actual social forces that could enact such a programme.

This book sets out the transition as a field of social and class struggle, outlining what political, economic and social transformations are being pursued under the guise of the shift to a low-carbon economy; how they relate to social conflict, and what the strategic and organisational implications are for workers, environmental campaigners and the Left. While much of the transition is a question of renewed rounds of enclosure and colonialism in the Global South, this is a book focused on the transition in the Global North. By Global North I mean wealthy, industrialised nation-states, primarily located in Europe, North America and parts of East Asia, that make use of military, political and/ or economic power to dominate and exploit the rest of the world.[1] This is because the 'Imperial Mode of Living'[2] in the North is the principal cause of climate change, and it is pro-duction and consumption in the Global North that must be radically changed if there is to be any hope of preserving a liveable climate. The war of transition in the Global South is qualitatively different as a political struggle, the more so as the transition is a profound reorganisation of the international order, one that affords innumerable opportunities for regional and national movements in the South.

This is not a book about what the transition should look like. It is not a perfect analysis of the new industrial paradigm of the transition economy, or a plan for when the Left finally seize control of government. It is an attempt to articulate the ways things are changing right now, and to outline what social forces already exist which are capable of intervening in the transition.

This book is about how we fight the war of transition – and win.

The Future of Port Talbot

On 11 November 2023, 400 steel workers and their families took to the streets in Port Talbot, south Wales, to protest the threat of job losses at the largest steel mill in Britain. The sky was bright as the workers and their families marched along the waterfront. Talking with the small number of journalists covering the protest, workers told of the terrible cost the job losses would have, on not just them but the local community.

One in six jobs in Port Talbot are in the mill. Four thousand people are employed at the mill, and 2,800 jobs are set to be cut when the blast furnaces are shut down. Port Talbot owners Tata Steel had threatened outright closure without more government support, citing a lack of profitability, high energy costs and competition from cheaper Chinese steel. Tata Steel is far from struggling financially, having paid £1.4 billion in dividends to their shareholders since 2019. Yet it is true that steel can be made more cheaply elsewhere, and that blast furnaces must be closed as a part of the effort to reduce carbon emissions. Tata Steel demanded support for a transition away from coal-fired blast furnaces to greener, lower-carbon electric arc furnaces. The government obliged, committing £500 million to the transition – around a third of the total cost.

Tata Steel then announced the job losses. The government spun the story in a different direction, declaring it a win for workers; saying it had 'saved' thousands of jobs through 'one of the biggest government support packages in history' as a part of the shift to a low-carbon economy.[3]

After the announcement of potential job losses, Unite the Union, representing over a quarter of the Port Talbot workers, put forward an alternative plan. Unite's plan was to keep using the blast furnaces until the end of their functional life – 2027 for one furnace, 2034 for the other – and to install an electric arc furnace alongside the blast furnaces while investigating ways to make the latter less polluting, possibly using alternative fuels like hydrogen. This plan didn't just maintain existing levels of steel production – it called for production to be doubled. Alongside this, Unite demanded the government invest in the entire region to turn it into a green industrial hub.

However, the union's plan always relied on the British government being something that it's not, and on the transition being a project of national renewal rather than a plan to revive economic growth through pro-market policies.

Since the 1970s, the steel industry across the Global North has been in decline. Port Talbot once employed 18,000 people, making it the single biggest employer in Wales. In 1971 there were 323,000 steel workers in Britain.

Now there are only 35,000. Similar falls have occurred across the Global North, and numbers are set to shrink further in the course of the green transition. As steel production is transformed into a green industry, fewer workers will be required, meaning that as the transition gathers pace the number of jobs lost and communities gutted by deindustrialisation will only grow.

Steel production is vital for the foundational technologies of our so-called green future: electric cars, wind turbines, solar panels and infrastructure projects such as rail links. Yet while it is essential for the green transition, it is also very much part of the problem. Steel production is responsible for 11 per cent of global carbon emissions. Port Talbot steel works is the single biggest carbon emitter in Britain. Switching from coal-fired blast furnaces to electric arc furnaces will shave 1.5 per cent off Britain's emissions in one stroke – a huge amount for such a site-specific change.

Electric arc furnaces use scrap metal as raw material. Despite claims to the contrary, the steel produced through this process is strong and durable, fit for almost any use. Thirty per cent of all steel is already produced this way, making steel the most recycled material in the world. And Britain produces more than enough scrap to supply all its needs, exporting more scrap than it uses in new steel each year.

The money Britain's government has committed to Port Talbot is part of a global return to strong national industrial policies. Billions have been allocated to the transition away from carbon-intensive processes and towards solutions for industries such as steel. Tax breaks, funding for research, direct grants and not-so-subtle adjustments to pricing and market policy – all these instruments and more have been deployed to underpin the transition towards a low-carbon economy. Despite this, however, the decarbonisation of steel is not happening nearly fast enough. The steel industry is not on track to halve emissions by 2030, or to fully decarbonise by 2050. In fact, total emissions from the sector are still rising, and there are significant plans to build new high-emission blast furnaces around the world.

Putting Food on the Kitchen Table

During the November 2023 march and in union meetings, Port Talbot's steel workers questioned how they would pay their mortgages and bills or put food on the table. We could just as easily have started this story there, on the kitchen table.

That kitchen table would have looked very different only a few years prior. Not because of any major shift in what people liked to eat, but because there would have been less on it. In Wales, as across the rest of Britain, people are buying less food. Four in ten people in Britain, and as many as six in ten in poorer areas, are cutting down. What food they buy tends to be cheaper, more highly processed and less nutritious. People have also cut back on fruit and vegetables as prices have increased, sometimes by huge percentages, while there has been a shocking rise in illnesses like scurvy and rickets, as well as a four-fold increase in malnutrition. This deterioration is not confined to Britain. In 2023 hunger continued its multi-year rise in the US, reaching its highest point in a decade, while Europe did not fare much better. Between 2022 and 2024, global prices for a wide range of foods exploded, rising between 15 and 30 per cent. Millions were driven into malnutrition and hunger. At the same time, shortages and disruptions to basic supplies, from wheat and rice to cooking oils and vegetables, became commonplace.

The year 2022 was one of extreme weather. Heatwaves and drought devastated crop yields in Britain, Spain, France and Italy, sparking fears that 'heatflation' had taken hold of Europe's food supplies. While drought parched Britain, Italy declared a state of emergency due to flooding, and France was lashed by storms. Flood waters also covered a third of Pakistan while parts of the United States, including California and the Midwest, the country's breadbasket, continued to bake under a twenty-two-year-long mega-drought. Floods destroyed 1 million hectares of farmland in West Africa, and a combination of drought, floods and powerful storms produced food shortages and famine along the east coast of Africa. Disaster continued into 2023, with countries responding to shortages with export bans on key crops such as rice, and price controls to ward off social unrest. Agriculture executives, farmers and industry analysts all concurred that this level of simultaneous agricultural disruption was unprecedented.

It wasn't just food that became more expensive and scarce. Energy prices shot up, while storms and flooding affected supply. Everything from sand and glass to critical minerals experienced shortages and added to soaring inflation. Mounting disruptions and disasters cost national economies billions, while millions of people lost their homes and were displaced.

All of this is taking place at a time when wages are largely stagnant or in real-terms decline. Across the board living standards have dropped, sometimes dramatically, as millions are un- or under-employed, forced to turn to charities and food banks, or go without heating or food. Meanwhile, economists warn that people will just have to get used to living with less as the price of adjusting to a new economic reality.

Between the Climate Squeeze and the Installation Economy

Port Talbot is a story that will be repeated. The emerging transition economy will steadily shed industrial jobs, while making, and breaking, vague promises of future employment. Broken promises, not new industries, are what we can expect. Broken promises form the emotional and discursive foundation of the transition. As the assurances stack up, industry by industry, then fail again and again to be fulfilled, what emerges is a political atmosphere thick with a sense of betrayal.

Across the Global North, the financial crisis of 2008 initiated a long period of anxiety and economic deterioration. Industry and government both tell us that thanks to the green transition, this crisis can be overcome and there will once more be a future for us. Climate change is not a threat but an opportunity, if we can only embrace the project of that transition.

The bleaker truth is that the transition will not revive our economic fortunes. It does not offer hope of a manufacturing revival or a return to strong economic growth. The transition economy is, at best, an installation economy. The work of installing solar panels or maintaining newly built green infrastructure, is, despite its manual nature, service work. And it is service work organised largely in and through small and medium-sized businesses: a far cry from the mass employment of corporate factories. Conditions and pay are poor, and firms survive on razor-thin margins. Installing products made elsewhere, or managing the downstream waste – that is the green work of the future. This is not to say all manufacturing will disappear, or that there will be no jobs in construction and engineering. It is to say that the hoped-for industrial revival is not coming anytime soon.

Green growth is supposed to renew the compact between capital and workers, leading to increased wealth for all. Instead, what we see is a deepening social crisis. Climate change is driving up prices and creating sustained disruptions to food supply. Coupled with the steady increase in the price of energy and other core materials, due to the impacts of both the climate crisis and the buildout of the transition economy, what we find is a steady squeeze on real incomes. Climate change plus the effort to transition to a low-carbon economy is making us poorer. And as prices go up and supplies fluctuate, most of us will become not just poorer but exhausted.

Climate change is creating an increasingly hostile working environment for millions of people. There has been a fivefold increase in workers' exposure to extreme heat globally over the past twenty years. If the cost to business has been huge, the cost to workers has been even higher. Hundreds of thousands of people have died due to extreme heat in recent decades, with agricultural, installation and construction workers faring worst. At the same time, rising temperatures are also a problem for workers labouring in buildings and factories not suited to changing climates. This affects not only the obvious candidates of factory labourers, but also teachers, nurses and warehouse workers, making exposure to heat a unifying workplace condition.

Besides heat stress, climate change will drive a seemingly endless series of disruptions. Homes will be flooded, businesses forced into bankruptcy. Essential services will become harder to access while everyday tasks become less certain and more anxiety-inducing. Climate change will worsen our health, exacerbating everything from diabetes to disease. The current mental health crisis across the Global North will intensify, making it harder not just to get by but to organise against the crushing pressure of climate change.

The demand for care will surge. Women, already doing the vast majority of waged care work, will be asked to do even more, under worsening conditions. At the same time, against a continuing backdrop of government austerity, the burden of the double shift, of caring for family and friends, will intensify. While the stories told of the losses faced in the green transition overwhelmingly feature men, it is women who will bear the brunt of the impact.

This is the squeeze. During the transition life will become meaner, diminished, as we all learn to live with less. And not just live with less but hope for less. As incomes fall and security disappears, our horizons will shrink. Hope, already battered by broken promises and failed investments, will dwindle along with our horizons.

This Doesn't Feel like We've Won

The world's major fossil fuel producers are all expanding production. The biggest oil and gas companies are projected to spend US$932 billion on new oil and gas fields by 2030 and US$1.5 trillion by 2040. From new coal mines and power stations to long-term deals on LNG supplies – coal, oil and gas all seem not only here to stay but unlikely to decline in a substantive way anytime soon. This planned production would not only kill the chance of keeping future global warming below 1.5°C stone-dead, it would also make it impossible to stop it at 2°C. By most accounts, fossil fuel production must start declining right now, almost completely ending by 2050, to have even a chance of staying under 2°C.[4] Even this strikes many as optimistic.

Instead, demand for coal, oil and gas reached record levels in 2023. While the International Energy Agency (IEA) suggests that demand for all three will peak by 2030, it is likely that oil and gas demand will plateau around then rather than start to decline, and stay high until at least 2050, in line with oil and gas company expectations. Alongside this, efforts to curb deforestation, reduce pollution and protect the world's oceans have all fallen far short of promises, and in some instances, such as deforestation, gone into reverse.

Across Europe, regulations and laws mandating the switch to heat pumps, electric cars and building insulation are all being weakened or pushed back to later dates. Trump has been re-elected to the White House, vowing to wind back government spending and support for green industrial programmes and climate commitments. Industry after industry is securing either delayed regulation or weaker targets and special exemptions. Countries are refusing to strengthen their already inadequate pledges to reduce carbon emissions, with many planning to renege on the existing ones. At the same time, grassroots and union opposition to climate measures is growing, joining the already existing anti-green campaigns run by business interests.

The gap between what's been promised and what is needed to arrest the climate crisis is huge. The brutal reality is that we are on track for around 3°C of global warming, possibly as soon as 2080. Given this, it would be fair to ask whether the transition to a low-carbon economy is really happening, whether talk of a green transition is just a lie and a distraction. Indeed, there is no shortage of people saying exactly that.

Yet for each critic there is a corresponding optimist, pointing out that, yes, progress is not rapid enough, but look! Look at the exponential growth of renewable energy, growth so fast and energy so cheap that we can see the end of fossil fuels much sooner than predicted. Even China is predicted to reduce its emissions faster than expected! The optimists point to the remarkable surge in electric car sales, the fact that almost all economic activity and most major corporations are covered by net zero targets and regulations. From the adoption of circular economy principles and policies to massive subsidies and take-up of heat pumps, they argue that a profound transformation really is underway and is happening more rapidly than many expected.

Green capitalism is no mirage. Just the listed green economy has an annual revenue of US$5 trillion.[5] Investment in Greentech will rise to US$2 trillion in 2024, double that of fossil fuels, according to the IEA.[6] As a sector – or fraction – of capital, it is growing faster than any other bar the tech sector.

The transition is happening, and yet it's not. It's both too slow and much faster than expected. We can begin to square this circle by understanding that the main purpose of the green transition isn't to keep future climate change to 1.5°C, or even 2°C. The function of the green transition is to turn the problem of climate change into a new capitalist frontier. And while some of the profiteering involved in cashing in on the climate crisis is deeply cynical, much more is pragmatic in its commitment. But we should be clear: the implementation of this solution is first and foremost driven by economic imperatives, not scientific ones. It's a means to restore the economic growth and fortunes of the national economies of the Global North.

The idea of using climate change to tackle social and economic problems dates back to the late 1990s, but the particular mix of government interventions and policies that characterise the transition economy first gained favour after the 2008 financial crisis. Since then, governments in the Global North have made some attempts to shape their domestic and regional economies to revive economic growth and spur both innovation and productivity. China started earlier, with by and large the same ends in view.

The transition economy is organised around making business feel secure in its investments, not the direct organisation of production by state industries. Government contents itself with encouraging and derisking investment. It mobilises legislation and policy to create new markets for green goods where prices are set to ensure reasonable profit margins. The aim of this conducive investment environment is to persuade business to build out the transition economy, pushing it forward on a wave of privately funded – and state-supported – innovation. Success is measured not only, or even primarily, by counting carbon reductions but by capital expenditure (capex) figures and how effectively these innovations drive growth rates.

It is an economy built on borrowed parts, many of them neoliberal. It is an economy that ruthlessly punishes the poor, enables the continued assent of asset managers and monopolists, encourages rentierism and calls for the relentless draining of resources from the Global South to build this shiny green high-tech future. All in the name of ending the iron grip of stagnation that is suffocating the economies of the Global North.

The War of Green Transition

Despite it being a very real social and environmental disaster, we can't simply oppose the green transition. But this doesn't mean reducing our political actions to symbolic protests against false solutions, or merely hoping for the best while voting for the lesser evil. There are two tasks ahead of us. We must leverage our power to block not only their transition but continued fossil fuel use. And we must do so while creating an alternative path to a sustainable social economy. The second task is the harder one by far: we have to accomplish it while building the movement we need.

Any alternative to their transition must start from the most basic of facts. Without a militant progressive environment movement capable of forcing an end to the fossil economy, there is no chance of making a different economy, let alone a better world.

The fossil economy is a vast web of infrastructure. It is refineries, pipelines and mines. It is not an abstract system beyond our reach. It exists in physical space, and that makes it vulnerable. If we are to disrupt this destructive carbon web, we need to do more than perform symbolic actions in city centres, or protest against government inaction. We need to physically put ourselves in the way. We need to blockade the fossil economy.

The blockade as a form of militant action disrupts the smooth operation of critical infrastructure, stopping flows and isolating key nodes within the fossil economy. But it is more than just a physical obstruction. It gains its effectiveness not from the disruption it causes, but because it is a machine that enables the disruption to endure. A blockade extends well beyond the protest site into a network of recruitment, training and support. It is a means by which we can create militants and sustain them in action.

Blockades have impacts far beyond the site of struggle. They exert pressure on business finances, enforce compliance and ultimately weaken the target, forcing change. They are a means of inflicting a war of attrition on fossil capital, wearing it down and making an energy transition away from oil, gas and coal a reality.

As our lives are crushed by the transition, as shelves grow bare and the cost of living increases, we have to organise against the climate squeeze. The green transition as a project continues the naturalisation of the market undertaken through neoliberal policies. Where neoliberal politicians worked to liberate business from public control, to weaken the threat of democracy to capitalism, the transition will push this project further. Climate change will be invoked to make markets seem beyond political demands and social control. It will be climate change that increases prices and makes us all poorer. But there is nothing natural about a price. Prices are political objects. We need to organise to contest them from below, to impose controls on necessities and fight against being made to pay not only for the climate crisis, but for the costs of the transition.

Finally, we must grab hold of the transition in order to transform it, to make it into something that meets our needs and our dreams. We can't fight the war of transition on the defensive. We can't resist their transition plans one workplace at a time, nor can we simply hang onto polluting industries or destructive jobs. There is no one workplace or industry that holds leverage over the entire economy. The transition is work – it is made piece by piece. But that work is not reducible to heavy industry or even waged work. If we are to take hold of the transition as a process, we need to go beyond the workplace as the privileged site of our demands for a just transition.

Waged work has always been embedded in place: it is always somewhere, bound up in a community. The closing of blast furnaces in Port Talbot means more than the loss of thousands of jobs – it's the gutting of a community. And not just in terms of income and post-industrial decline, but of the very sense that a future exists to be had there. What is being transformed are communities. Therefore, the focus of our organising efforts has to exceed any one workplace, and start within the community.

Approaching the community as a single site, drawing on both workplace and community-organising tools, opens up the possibility of building a broader struggle, one that can be anchored in local institutions and leverage the power of a mass, community-based movement. It also creates the possibility of embedding workplaces within a community, and asking how they can serve our needs, rather than those of business. Putting our needs and those of the community at the centre of our organising makes it possible to move from defensive struggles to campaigns that aim to recreate the economy for our own ends. Whole community organising in the transition means making counter-plans, plans that set out not just our opposition, but what we want our lives to look like.

The seemingly inevitable decline of the steel mill in Port Talbot is both the emerging reality of the green transition economy and a symptom of the impasse in which progressive and environmental movements find themselves. The Left ought to have answers beyond defending polluting industries, and environmentalists should have more to say than platitudes about a 'just' transition. Port Talbot is a story being repeated around the globe, and we witness a growing sense of betrayal as the promises of the green transition turn out to be all too obvious lies.

We are in the midst of a profound moment of transformation. Hegemony is yet to be established, and old norms and rules are breaking down. This should be a time of feverish activity and organising. We should be pushing forwards, building our power and imposing ourselves on the transition. The stakes of this war of transition are nothing less than a future worth living in.

The Squeeze

It's shocking just how quickly having to choose who eats breakfast, or whether it's more important this week to spend on food or on heating, has become normalised. In 2023 in the US, one in three parents skipped meals to feed their children, while one in four parents did so in Britain.[1] Millions of people in both countries skipped meals regularly because they couldn't afford food.

Food prices increased across the US by almost 6 per cent in 2023, while British food inflation peaked at 19 per cent in March of that year. Britain faced food shortages as supply was hit by storm damage and bad weather. Shelves in supermarkets were bare of tomatoes and salad veg, with restrictions on how many cauliflowers and broccoli shoppers could buy. In 2023, bad weather hit potato, leek, carrot, barley and wheat crops, and olive oil prices skyrocketed as drought punished Spain. At the same time, between 2020 and 2023 British households saw their energy bills increase by 80 per cent and their water bills by 19 per cent, while the cost of most goods and services rose by around 10 per cent.

In the US there were few shortages. Instead, food inflation led to a surge in food insecurity as 47 million Americans struggled at least some of the time to obtain enough food.[2] Inflation rates fell in the US from their 2022 highs, but were still significant enough to erode living standards, while leading to a jump in both consumer debt and filings for bankruptcy.

Across the world over the last four decades, living standards have broadly declined as real wages stagnated and welfare states were hollowed out. Malnutrition and surging physical and mental health crises have shortened life expectancies in the US and Britain. When coupled with the squeeze on incomes and rising inequality, what has emerged as an increasingly common condition is what has been called 'shit life syndrome': a situation where a person's life is so consistently difficult and stressful due to the convergence of social and economic crises that it produces profound mental and physical health problems, which in turn compound poverty and social isolation.[3]

In 2023, the Bank of England's chief economist, Huw Pill, announced that, rather than fight for higher wages or demand functional government services, people would just 'need to accept' that going forward they were going to be poorer.[4] Of course, it's unlikely that becoming poorer is an outcome Huw Pill himself will need to accept. His £180,000 per year salary no doubt keeps him well supplied with olive oil, while millions of others are struggling to afford the bare necessities.

While central bankers heroically tackled inflation by making us (even) poorer, economists declared the surge in prices temporary, repeatedly ascribing it to the Russian invasion of Ukraine. Journalists equivocated, agreeing that this could be the impact of energy price rises, or else – citing the now-clichéd litany of ecological catastrophes – it could be a harbinger of worse to come as the climate crisis accelerates.

Most of us are all too familiar with the ominous warnings that fill our news feeds. Every day it transpires that things are, in fact, much worse than we had originally been told they would be. Against the brittle professional optimism of policy-minded climate campaigners, the prevailing senti-ment among climate scientists and researchers is one of profound anxiety, not about the future in ten, twenty or fifty years but about the impacts the climate crisis is having right now.

We already know climatic change is making a drier, hun-grier, poorer world. We also know that the transition is likely to disrupt what precious stability remains in our lives: jobs have been and will continue to be lost, communities are under threat and the costs to us keep on mounting. And while the green transition has most definitely begun, it will not happen fast enough or deeply enough to stop the changes to the climate that will profoundly transform the world in which we live.

War, inflation, drought, floods – these are not portents of what might happen if we fail to 'act now'; they are not omens. They are in fact symptoms of how the climate crisis and the broader unravelling of the biosphere are fundamen-tally transforming the very nature of politics. It is these realities – of war, catastrophe and deprivation – that make clear the contraction and disruption of our lives that is already underway. Ecological catastrophe compresses our lives, reduces the space for action and drains the colour and richness from existence, making the remaining possibilities much thinner and more brittle.

This is the squeeze. The reality of having to choose who eats and who doesn't diminishes us. What gets smaller is our sense of the kinds of lives we can lead, as even what we can hope for shrinks.

At the same time, political action is not energised by the mounting impacts of the climate crisis, it is crushed by them. It's not 'doomism' that disables and extinguishes political action – it is the material reality of the catastrophe itself. Politics, like everyday life, is shaped by what we can imagine, and is bound to the material world. As material possibilities contract, and disruption and crisis come to characterise daily life, political action itself becomes a casualty of the climate crisis, degraded along with wheat yields and freshwater supplies.

The squeeze is not just a description of a series of unfortunate events, nor is it a list of the impacts of the climate crisis. The squeeze enables us to name and take hold of the new terrain of political action. Everything – from the price of eggs to which essential service will be closed – will for the foreseeable future be shaped by the squeeze. The squeeze will animate political campaigns, government policy and business strategies. This is because the steady downward pressure, punctuated by a series of all-too-predictable disasters, on living standards and life expectations, will not end anytime soon. Things getting worse, dramatically worse, is now a near-permanent, long-term feature of the global economy.

The squeeze is what the climate crisis feels like from our point of view. We can start to trace the mechanisms of the squeeze through the foundations of social and political life, what Raj Patel and Jason Moore call essential 'cheap things' – food, energy and people.[5] These 'things' are vital resources that must be rendered cheap, whether through destructive practices, violence, colonialism or exploitation, in order to enable capitalism to function and to ensure business is profitable.

The squeeze is a specific historical moment where the strategy of never-ending expansion has become exhausted, terminating in both the climate crisis and the transition to another kind of 'green' economy. The problem for us is that from capital's point of view, it is a dizzy moment of both promise and dire threat, one that holds the potential for a renewed cycle of profiteering and sectoral growth and, at the same time, poses real risks to investments and entire industries. But for us, this moment is characterised by contraction and disruption to food and energy supplies, entailing risks to our health and our lives. The squeeze is transforming the grounds of the everyday, and our ability to act on the world and shape the future.

Peak Rice and No More Olive Oil

In September 2023, for the second year in a row, Europe almost ran out of olive oil. Drought, fire and storms devastated olive crops, driving Spanish harvests down 40 per cent, with yields down 80 per cent in some regions. Spain produces 50 per cent of the global olive harvest, and accounts for 70 per cent of the European Union's supply. The climate crisis is rarely confined to a single national geography. Greece's olive oil crop was down by half, and Italy and Portugal also suffered poor harvests. Globally, production fell short of demand, with only 2.4 million tonnes produced against demand for 3 million tonnes. Prices skyrocketed, farmers went bankrupt and newspapers ran scare stories about the end of olive oil. After 6,000 years of cultivation, climate change has pushed olive production into a state of profound crisis.

Cooking oils – from sunflower oil to palm oil – contribute 10 to 20 per cent of people's daily calories globally. Any crisis in the production of cooking oils is significant. The major two crops that contribute to the production of cooking oils are palm oil and soybeans. Harvests of both have suffered in recent years due to extreme weather events and temperatures. Brazil and Argentina, the number one and number three global producers of soy, have seen significant crop losses.

Palm oil production is a major cause of deforestation and is profoundly unsustainable. As a commodity, palm oil is found in everything from cooking oils to ultra-processed foods to industrial lubricants. Any shortages or disruptions to supply drive up costs, not only to cooking oils but to a much wider range of products. The leading palm oil producers, Indonesia and Malaysia, responsible for 80 per cent of the global crop, have both suffered crop losses and reduced harvests in recent years. Production costs are rising in both countries, reducing output and increasing prices. Palm oil was more expensive in 2022 than at any point in the past fifty years.

Cooking oil is far from the only staple crop to have suffered from climate-related reductions. Over 2022–23 most major agricultural crops experienced losses, including sugar, wheat, fruit and rice. Pressure on rice production was so extreme that the *Economist* ran a series of articles exploring the idea that we may have reached 'peak rice'. Other headlines foresaw the end of chocolate and coffee as climate disasters struck both, while increasing food prices have led to profound changes to consumer behaviour and a more generalised reduction in the amount of food being bought.

The media headlines and the prophetic warnings of environmental activists often make it sound as though food is running out, and we all face a future of famine. But it's more accurate to understand agriculture as a specific industrial system being pushed into crisis. This food crisis manifests as a series of contractions. Yields are reduced while nutrient levels drop, resulting in less food, of lower quality. Arable land is lost to flooding and drought. Higher temperatures limit the number of days workers can be safely outside, reducing the available time to harvest. Diseases spread, while water supplies drop. These contractions and disruptions are not evenly distributed and are punctuated by regular disasters, reflected on supermarket shelves as 'out of stock' signs. Contraction is not only a question of less, and disruption not only a matter of supply. The squeeze is also a question of how much money we have to spend to eat. While the total absence of cooking oil, or bread, cannot be solved by money alone, for the most part hunger is an economic condition. The squeeze signals a period where food prices will climb, competing with other costs. For those in the Global North this will undermine the few remaining post-war social gains; in the Global South, where a much greater proportion of household income is spent on food, malnutrition and hunger are already surging and will continue to do so.

The climate crisis has already reduced yields by 10 per cent in four major crops, and it's been estimated that global farming is 21 per cent less productive because of climate change.[6] From here on, the food crisis will only worsen. It's estimated that at a global level food production will decline by as much as 14 per cent by 2050, adding over 500 million people to the already huge number facing malnutrition and hunger.[7] As the world lurches towards 2°C of global warming – likely a brief pit stop on our way to 3°C or even higher temperatures – agriculture and food production will be increasingly curtailed. According to best estimates, Earth will reach 2°C of global warming around 2040. At this temperature global crop yields will decrease between 5 and 20 per cent.

Water availability will decrease even as extreme flooding events and storms devastate crops, further depressing yields. At the same time, reduced water supply will add to the cost of living as efforts to secure enough water, including desalination and improved water infrastructure, will drive up costs to consumers. Water prices are already increasing faster than inflation in many countries, all while private investors, cashing in on neoliberal policies, make millions.

One of the effects of global warming is a rise in insect attacks on crops, leading to devastating locust plagues such as the one in Pakistan in 2019 that cut 2 million tonnes from agricultural yields and forced the government to import wheat for the first time in a decade. With a warmer climate comes a rise in pathogens, too. The spread of fungal infections in crops alone could cut upwards of 13 per cent of global grain production.

All of this speaks to a squeeze on food availability and increasing prices. Yet the squeeze is not just an effect of constant pressure but the result of a series of shocks and disruptions. Severe flooding, wildfires and record-breaking storms have all smashed harvests in recent years. While we could read this as yet another factor depressing crop yields – which is true – it is better to think of these events as profound shocks that our globalised food system can't withstand. As storms wipe out wheat harvests, or fires devour olive groves, the sudden loss of supply upends our just-in-time minimum-stock food systems, leaving shelves barren and consumers desperate, while at the same time enabling commodity traders to drive up prices even further, profiting from speculation on just how hungry we will get. Such extreme weather events already occur not only more frequently, but often in quick succession, making recovery impossible. Increasingly they also happen around the world simultaneously, putting food security in jeopardy. That several of the world's breadbaskets might collapse at the same time, risking global famine, is no longer a dystopian fantasy but a genuine risk.

This combination of contraction and disruption signals the end to the historically low food prices we have grown used to for the past few decades. Those low prices were the result of a vast, globe-spanning reorganisation of nature undertaken through the so-called Green Revolution and the new enclosures of the neoliberal era. The proportion of people's incomes that will have to be spent on food will now increase, while food security will decrease. In the Global North, where food accounts for between 5 and 15 per cent of consumer spending, this will manifest as a further squeeze on household incomes, driving some into deeper poverty. In the Global South, where food expenditure in relation to household income is already much higher, ranging up to 60 per cent in some countries, these increases and disruptions are already producing social catastrophes and sparking civil conflicts.

These changes will produce cascade effects reducing disposable incomes and other discretionary spending, both of which have already been under pressure since 2008, while increasing absolute poverty. This will reverberate through local economies leading to lower economic activity and growth, further un- and under-employment, the closure of small businesses and the collapse of local communities. What begins as a squeeze on food will quickly become a crushing economic condition.

Blackouts and Failing Grids

Alongside food, energy has been one of the main drivers of the squeeze; and as with food, the rule of thumb is that the poorer you are, the bigger the impact of rising energy prices.

It's trite to point out that energy is the foundation of social life. Despite the fact that the green energy transition is well underway, fossil fuels still make up 81.5 per cent of our total global energy supply. The International Energy Association predicts that fossil fuels will still supply 20 per cent of global energy by 2050, the supposed deadline for reaching 'net zero'.

Yet it is often unappreciated quite how fundamental energy is to economic stability and the actual cost of everything. Around 20 per cent of food prices arise from energy costs, meaning a 1 per cent increase in oil prices translates into a 0.2 per cent increase in the price of food. More broadly, everything from plastics to clothing to medicine either contains or relies on fossil fuels, so that every time oil and gas prices jump, so too do consumer prices. Shipping, transport and logistics all depend on oil, as does aviation. During 'normal' periods, oil prices contribute around 10 per cent of base inflation rates. During periods of volatility and crisis, oil and gas prices contribute upwards of 40 per cent of inflation rates. At the household level, energy contributes around 15–20 per cent of direct costs and impacts a further 45 per cent of household spending indirectly.

The cost of oil and gas is not just a factor of war and crisis. While talk of peak oil has long since gone quiet, the fact is that producing oil and gas is getting more expensive. The economic cost of producing a barrel of oil (or its gas equivalent) has been increasing for years now and is expected to keep rising, as conventional sources reach the peak of their production and extraction turns to more 'unconventional sources'. Unconventional sources, such as shale or tar sands, are more expensive to work, and so require higher sale prices to break even. Equally important is the energy cost involved. Extracting oil or gas isn't just a question of money but energy – it takes energy to pump oil out of the ground or turn tar sands into usable fuel. And it is becoming more energy intensive to generate fossil fuels, meaning we are using more energy each year to extract oil, reducing the efficiency of the extraction process and further driving up costs.

While the cost and stability of supply of energy remains crucial to the constitution of the squeeze, the impacts of climate change on energy production itself are just as important and deeply underappreciated.

Hot weather in July 2023 led to significant reductions in output in several French nuclear power plants. Nuclear plants need huge volumes of water to enable them to cool sufficiently. It's for this reason they are usually situated on rivers or by the sea, where they are also at greater risk of flood damage and sea level rise. Coastal nuclear plants already shut down in severe weather. On the other hand, decreased river flows during the summer will reduce the amount of time nuclear power plants can operate. And we are already seeing significantly decreased river flows around the world. As climate change intensifies the power and frequency of storms, disruption will become a far more common occurrence. And this is not taking sea level rise and the flooding of nuclear reactors and waste storage into account.

Hydropower is the main source of energy in a significant number of countries – twenty-eight countries in the Global South use hydropower for the majority of their electricity supply, and 150 countries generate some of their electricity this way. Drought in the US has prompted water managers on the Colorado River to warn that lack of water could cause the Glen Canyon Dam hydrogenator to shut down, cutting power to 6 million people and seven states. The Kariba Dam, which normally provides half of the electricity used by both Zambia and Zimbabwe, is currently at its lowest level in history, leading to extreme power cuts in Zimbabwe of nineteen hours per day, and rationing in Zambia. The cuts have knock-on effects, including cutting water supplies (the pumps need power), and prompting regional governments to turn to coal as an energy source.

Across the board energy generation will suffer, and power outages are already increasing in frequency due to climate change. Thermal generation (coal, gas and other fossil fuels), like nuclear power, requires cooling systems. These usually use ambient air or water. As both get hotter, thermal power stations become unable to run their cooling systems, meaning they must shut down or run at a reduced rate. As the climate warms, thermal power generation will be reduced – it's already 1 per cent lower and will drop a further 1 per cent per degree centigrade on average. This will reduce supply and, due to heat stress, increase operating costs by around US$47 billion per year, ultimately driving up prices.

Extreme and unpredictable weather will reduce renewable energy production as well as damage vital energy infrastructure, producing shortages and further driving up prices. Storms and high winds already account for 80 per cent of all power cuts in the United States.[8]

The contraction of energy supply, be it from reduced capacity, disruption or supply shortages, will directly increase the cost of everyday life. These cost hikes and disruptions will cascade throughout the economy, driving bankruptcies and closures across small and medium businesses, pushing families into poverty and homelessness, further exacerbating the strain on local government services and finances, and impacting critical infrastructure from water treatment to telecommunications. It will directly increase the costs of transport, both public and private, electric and fossil-fuelled. It will further raise the cost of food and everyday items. It will increase fuel poverty, while decreasing people's disposable incomes. Along with the squeeze from rising food prices, this reduction in disposable incomes will impinge in turn on economic activity, triggering economic stagnation, class conflict and government austerity.

Deteriorating Health and Broken Hearts

The squeeze is not only a problem of prices and shortages. It is a question of deteriorating physical and mental health as social and health services are pushed to breaking point, amid constant disruptions.

If we count the climate costs to healthcare in monetary terms, the scale is huge – between US$2 and 4 billion per year by 2030. The societal costs of mental disorders provoked by climate change could reach US$47 billion per year.

But while projected figures like these can help us grasp the scale of the impacts, they miss crucial details of what it means to live them; to understand what such an experience does to us physically, emotionally and mentally. To understand what damage it does to our ability to care for ourselves and others, what it does to our politics and sense of hope.

In 2023, smoke from unprecedented Canadian wildfires covered an area from Minnesota to South Carolina, subjecting 55 million Americans to dangerous levels of air pollution, and briefly making New York one of the most polluted cities on Earth. The fires, covering an area twice the size of Portugal, turned New York City's sky a dystopian orange, as hospitalisations due to the pollution spiked. Fires in Indonesia led to similar scenes in Malaysia and Singapore, forcing schools to close.

While wildfires might grab headlines, climate change worsens air pollution in other ways. It encourages the formation of ground-level ozone, usually known as smog. Smog forms when pollutants interact chemically with sunlight, and higher temperatures speed up these reactions, producing higher levels of smog. Alongside this, changes to weather patterns can lead to events like weather inversions, where a layer of warm air moves in over a region, acting like a lid or cap over the cooler air near the Earth's surface. This traps pollution closer to the ground, preventing it from dissipating and worsening air quality.

Air pollution has a huge range of impacts on our health. It is linked to respiratory and cardiovascular issues – from asthma and bronchitis to heart attacks and strokes. It can reduce lung function and lead to reproductive health issues. It increases the risk of cancer, kidney disease and diabetes, impairs cognitive function and neurological development, exacerbates allergies, skin conditions and epilepsy, and has been linked to various mental health issues such as depression, anxiety and cognitive decline. It is also responsible for 7 million deaths globally each year.

As in the case of wildfire air pollution, we tend to think of big, headline-grabbing events when we think about the health impacts of climate change. And yet, just as important are the slow forms of violence that constrict and damage our lives. Heatwaves encapsulate this kind of climate attrition. The summer of 2023 was Europe's deadliest on record, and one of the warmest summers ever recorded worldwide. That year, 61,000 people died in Europe, and thousands more around the world. Recent studies have set out the deadly cost of heatwaves since 1991, when climate change first became a global political issue, and the level of warming was already around 0.4°C above the pre-industrial average. Over 7 million people have died in heatwaves caused by climate change. As with most other climate impacts, the poorer you are, the more vulnerable – both within countries and between them.

Heatwaves overwhelm the body, spiking its internal temperature. As the temperature rises, our bodies sweat in an attempt to cool us down. As we sweat, we lose water. This dehydration weakens our ability to think clearly, thickens our blood and compromises our kidneys. Our skin gets hotter, putting our cardiovascular system under strain which, along with thicker blood, may trigger heart attacks and strokes. As our bodies heat up, our organs can't function properly and in some cases can begin to fail.

The heat is oppressive. Living under heat extremes is disabling as well as deadly. Aggression and violence both rise with the temperature. Irritability increases, sleep quality decreases. Suicide rates go up. And as the Earth warms further, this oppressive atmosphere will worsen. Heat, like pollution, attacks our very ability to be social, and to be political. And, of course, it devastates the economy. In 2022 alone, 490 billion labour hours were lost to heat stress. The same factor wiped out 4 per cent of Africa's GDP that year.

Each kind of disaster has its own shadow. Fires kill, but they also produce air pollution crises, as well as a series of illnesses, besides trauma. Flooding prompts outbreaks of disease, alongside profound levels of mental illness. Both can lead to spikes in homelessness and displacement. Millions are already displaced each year due to environmental disasters – over 30 million globally in 2023, including over 2.2 million in the United States. Billions more are predicted to be exposed to environmental disasters by 2050. Alongside the death, ill health and displacement comes the destruction of community and a loss of place, deepening the affective cost of climate change.

And then there are other impacts. The geographical ranges of diseases like malaria spread, putting millions at risk. Between 2030 and 2050, our near future, climate change will cause 250,000 additional deaths per year, pointing towards 83 million excess deaths by 2100. And this is probably an underestimate. What this estimate doesn't do is describe what it will feel like to live with the stress of increased air pollution, the anxiety of not knowing how bad the storm or flooding will be, the oppression of a monstrous heatwave, or the deterioration of our health as heat, disease and poor nutrition all take their toll on us and those we care for.

All of this adds to the immense emotional and psychological burden climate change is already placing on us. While austerity, economic stagnation, increasing precarity and poverty are all devastating, climate change too is driving a profound shift in our mental health.

Flooding, like that in Lewes, England, in 2000, or in Lismore in Eastern Australia in 2022, has critical effects on people's psyches. Beyond the impact of the immediate event, trauma including PTSD lasts for years. With the destruction of homes comes a sense of not only loss, but of pervasive insecurity. So too with fires and storms. Disasters linger long after the event as trauma is written into our bodies, eroding our sense of safety. They can destroy our sense of place, making our homes and communities unfamiliar and threatening, leading to feelings of solastalgia, the grief or existential distress felt for a lost place. Coupled with sleep lost due to rising temperatures and heatwaves, as well as the compound effects of reduced incomes and food and energy insecurity, plus pervasive disruptions to services and everyday practicalities, life will be – and already is – far more stressful and challenging, even as disruption becomes a stupefyingly ordinary event.

Of course, there is a real material cost to disasters, one that is crushing to households and businesses alike. From the mundane costs associated with repairs to those accrued from businesses shutting down, or days lost to work, all the way to worst-case scenarios where homes and lives are lost, storms and disasters extract a huge financial toll. Globally extreme weather cost the world US$16 million per day in 2023. The costs are increasing year on year – there has been a five-fold increase in the number of extreme weather events since 1970, with the damage done by each event increasing by 77 per cent.

It can be difficult to measure the impact on households. Often we have to use insurance claims as a proxy. In 2023 in the US, weather-related damage cost almost US$93 billion, while in Britain households claimed £573 million in 2023 to pay for weather-related damage – a 36 per cent increase on the previous year. British businesses also took a hammering, incurring £443 million in losses. These are the direct costs incurred, not the knock-on impacts. The average time it takes for a business to get back on its feet after flood damage is around fifty days. And given how many small businesses are run on thin margins, it's not uncommon for disasters to wipe them out, taking their jobs and services to the local area with them. In Britain, of those small businesses damaged by flooding, 40 per cent of them will permanently close.

Only two-thirds of households have insurance, meaning storm or flood damage could be personally and financially catastrophic. This is not just a question of the rising costs of premiums. More and more, insurance companies are deciding that particular kinds of events, or entire areas, are too risky to insure. The withdrawal of insurance extends to business as well. In 2024 two insurance companies in the US withdrew renewals and policies for tens of thousands of homes in Florida and California, while insurance company CEOs warned that the industry itself was at risk.[9]

The withdrawal of the possibility of insurance is already creating 'flood risk ghettos', while the same is happening for areas at growing risk of fire. Increasingly, areas will be depopulated as the costs of climate change mount. This won't be an even process. Those who can afford to move, will. Businesses will withdraw from the area, citing exposure. Insurance company CEOs are already telling people that 'they may have to move . . . that is the reality of climate change.'[10] The economic base of whole towns and regions will wither, hollowing them out in a way analogous to the worst impacts of deindustrialisation. Only those too poor to relocate, or tied to assets that will only continue to decline in value, will stay. The worst-case scenario for many in the Global North isn't being forced to move but being unable to move. In those instances, families could lose everything.

Recent years have seen a body of research into what's called ecological grief or climate anxiety. Across ten countries around the world, researchers found that an average 60 per cent of children and young adults were very or extremely worried about climate change, with 84 per cent describing themselves worried. Over half said climate change made them feel sad, angry, anxious, helpless or guilty. And 45 per cent said it negatively affected their ability to get on with their daily life.[11] This will come as no surprise. We all know – or are – someone whose life feels as though it is being stifled by fears associated with a warmer future. While professionally optimistic journalists and campaigners exhort us to not give in to despair, it's hard to escape the crushing sense that, if things are already hard, they are going to get much, much harder.

Similar numbers have been reported around the world for most countries, and most people – older, white, middle-class Boomers being the regular exception. Two-thirds of Brits feel worried, and half feel helpless. In Brazil the figure is 86 per cent, in India it's 80 per cent, and in the Philippines amongst young adults it's a staggering 92 per cent.[12] On average, three out of every four people in most countries see climate change as a major threat. Of course, it's not evenly spread. There are disparities between countries, though fewer than you might think and fewer every year. More substantive are the differences between generations, and between those identifying politically as either Left or Right.

The result of the squeeze on people for capital will be a shortage of labour. Combined with ageing populations and the breakdown of education programs, generating skill shortages, we can expect few things to work, and everyday life to be shaped by announcements of service disruptions and featuring 'closed' signs. This will be exacerbated by surging right-wing anti-immigrant politics, at a time when migration will only rise as climate displacement becomes a more profound phenomenon.

Worse still, this will be compounded by the fiscal crises of the state as a provider of services, already struggling under the weight of costly disasters. As the economic space within which people live shrinks, and services become ever harder to access and generally inadequate, expectations and hopes will also contract.

The Scarcity of Hope and the Contraction of Politics

What does it mean for hope to contract? Hope and expectations are relationships not just to the future but to what is possible. We hope for a better job, or that the government will stop the steel mill from being closed down, because we think it's possible that this will happen. What we believe we can expect, or what we believe is possible, gives hope its form.

Hopes are more than simple expectations, though. Hope is a mobilising and organising force that structures the direction of our lives. As memory shapes our understanding of the past and how we understand ourselves now, hope shapes our understanding of the future – who we could be, what our lives could look like. Both hope and memory give form and purpose to our actions; they give our lives meaning.

Hope is not unstructured – we are able to hope for some things and not others; or rather, we grow up learning to hope for some things and not others. Hope is a social phenomenon, produced through social forms and institutions – family, school, university, clubs, societies, political organisations and the media – and is endlessly circulated and mobilised to shape our social world. While ideology is a system of beliefs, values and ideas that describe the broader social and cultural atmosphere that frames our lives, hopes are the specific images and dreams that populate those frameworks. Our hopes orientate us towards some actions and goals and away from others. And just as ruling elites work hard to present certain beliefs and values as common sense, so too are some hopes promoted as worthwhile, while others are denigrated or dismissed.

Different societies produce different kinds of hopes. In fact, every single society produces a variety of kinds of hopes. Hope, like everything else, is political and contested. Rebellions and revolutions generate and circulate hopes, hopes that are repressed and materially attacked by governments and bosses, just as reactionary politicians and far-right activists likewise try to mobilise fears and insecurities in order to generate reactionary hopes and expectations.

Yet while there are always competing hopes and expectations, as with beliefs and ideas, some hopes become hegemonic. The hegemonic form of hope in the Global North after decades of neoliberal politics is aspiration. Not aspiration in the sense of aspiring to greatness in some heroic Greek sense, or something romantic and colourful. No, for those of us living in the Global North aspiration has a particular hue and tint – it means social mobility, envisaged as always more. It means a better job, more money, more things. This hope can be incredibly toxic – it tends to drive corruption, exploitation and the narcissism of entrepreneurialism. Or it is cruel, keeping us tethered to aspirational dreams that cannot and never will be realised.[13] Hope can be articulated differently, however, against these cruel and toxic dreams – we hope for more for our children, or for even modest comforts, in ways that aren't harmful or ultimately self-destructive. We also hope for a better world, and an end to the climate crisis.

For all the diversity of dreams and hopes, our aspirations are now tied to the climate crisis. More stuff means more carbon, more deforestation, more pollution. All the elements of a good life, all of the things that populate our happiness or dreams, come with high carbon costs. From new cars to gaming consoles to summer holidays on the beach – none are possible without producing climatic change. The foods we consume ravage the earth, while just scrolling on our phones uses enough electricity to destabilise the atmosphere.

This is not a moral claim, and it bears noting that lifestyle politics will not end the climate crisis. Our consumption is built on fossil foundations, not through our choice of purchases but by the actions of corporate tycoons and governments to maximise profit margins, plunder the Global South, maintain workplace control and ensure economic growth. These foundations were also built to overcome worker resistance and undermine ordinary people's power and autonomy.[14] Yet these facts do not make our consumer choices any less environmentally destructive. Nor do they make consumption in the Global North any less imperial or colonial. Our everyday acts and purchases depend on an unequal exchange with the countries of the Global South, draining resources and wealth and relying on both financial 'discipline' (the threat of economic punishments) and outright military intervention.

All of this is true, from our most mundane purchases – bananas and coffee – to our most fevered dreams of luxury cars or beach holidays in tropical paradises. The squeeze is felt so deeply because all our expectations and hopes are embedded in the climate crisis and the fossil economy.

Yet it's never been the case that everyone has the same access to the future – that we can all realistically hope for the same things. If our future is an expression of our present access to material wealth, then the future is just as inequitable.

Politics has long been concerned with overcoming the present in order to realise another future that already exists in the possibilities of the present, suppressed or hidden by capitalism. It is a promise of eventual victory, saying that though we might fail now there is always time to try again. With the creeping impacts of the climate crisis, hope is not only becoming scarcer and more ill-distributed, it is also changing. The futures that our present political moment let us imagine are contracting and becoming smaller, meaner. The idea that there is always a next time for revolution seems thin; the feeling that there is still the potential for a better future, one of collective wealth, is receding.

And while this is logically not the case – we know there are other kinds of wealth, other kinds of social and economic thriving – our lives, expectations and hopes are so bound up with existing forms of material wealth that we struggle to invest in these counter-hopes. The erosion of visible alternatives to capitalist realism over the course of neoliberalism has made it all but impossible to see past a specific idea of wealth and mode of living. As awareness of the climate crisis has grown, people have begun to imagine their futures with anxiety. Most of today's teenagers assume that their lives will be worse than their parents', and their grandchildren's worse still. People increasingly see the world as a zero-sum game – for them to get ahead, someone else has to lose out.

The climate squeeze, then, is very much a contraction of the promise of the future. As the material basis for our hopes contract, as disruption becomes the background to our lives and instability the norm, the future available to us shrinks. Paraphrasing the Bank of England's chief economist, people will just have to get used to hoping for less.

This squeezing of the future also reduces the scope for political organisation and action. Politics is a physical activity – it's something you do with other people. The time it takes to participate in political activity and activism and the requisite social skills (like knowing how to participate or the confidence to volunteer) have always been unevenly distributed along racial, gendered and class lines. And the squeeze is making this worse. People live more chaotic and disrupted lives, with less time to spare. Life is more atomised, and our sense of belonging has narrowed. The squeeze contracts our very ability to be political, to organise against the contraction of our lives, thus making political activism even less accessible and even more of an expression of social privilege.

But the future is squeezed in another way. The climate crisis, and the reality of what it means in practice to undertake a green transition, reduce the range of possibilities we have access to.

In some cases this is obvious. From what we can eat, to the places we can live, to the very certainty we need to make plans. But it also extends to the work we can do and what political possibilities exist that we can fight for. This is clear when talking about how workers can combat redundancies or closures in stranded industries – coal mines or fossil-fuelled cars are the go-to examples. It is less clear when it comes to those industries like steel that are necessary for the transition yet will still be radically transformed, resulting in less work and hence fewer workers. How do you fight for jobs or against closures when the future of steel involves fewer workers, and when there is already so much surplus capacity? What possibilities exist for the communities surrounding the steel mills?

The same question plays out right across the transition economy. While this might be booming in some countries, the job opportunities are not evenly distributed. And they aren't a one-for-one exchange with existing forms of work. Recent graduates can find professional work in a Greentech firm in London or Los Angeles, whereas an oil worker faces a more uncertain future. Even more uncertain is the fate of the communities that rely on these industries. Maybe government and business will invest in new industries, and maybe these communities will become centres of renewable energy manufacture, or Greentech business hubs; but it's more likely that as the oil and gas industry winds down, far fewer equivalent jobs materialise, and we see a profound deepening of the deindustrialisation of the past decades. While the green transition might promise some 'good' work, the reality of the transition suggests a future at least as poorly paid and precarious as the world we have now.

As with Port Talbot and Tata Steel, when it comes to work and the green transition, what can be fought for and won, what can even be considered as 'realistic' to hope for, is as much a part of a bigger story – one of international competition and stagnant economic growth – as it is a matter of storms, floods and the need to reduce carbon emissions.

2

Working the Transition

'No one knows what a green job is . . . People just want work. They want to know how are they going to earn a decent wage [sic].'

Jeeger Kakkad, industrial policy expert,
quoted in the *Financial Times*

Early in 2024, not long after the news of Port Talbot broke, I was on a call with union organisers for British Steel as part of an advisory group. British Steel had started to develop their own plans for a transition to electric arc furnaces, and the union organisers were seeking information on how they could avoid the fate of Tata's Port Talbot workers. The first question they asked was what technology they could demand to be implemented that would save jobs. We had to tell them that there wasn't one.

There isn't one because green steel doesn't need as many workers to produce the same amount of steel. For most manufacturing, green means less labour.

Transition doesn't inherently mean fewer jobs. It could mean fewer hours worked per employee. Or it could mean government investment in the local area to create new jobs.

But the work of building the transition doesn't automatically involve the creation of new, better jobs. There is no teleology bound to green frontier industries and technologies.

The green transition has been presented as a win for both workers and the environment: this is the promise that underpins its tenuous legitimacy as public policy. In its most radical incarnation, that of the Green New Deal, the transition to a low-carbon economy would benefit everyone. Supposedly. Capitalists would get government support for new investments; government would revitalise industry and ensure economic growth while stopping climate change; and workers would get good, secure jobs as well as a stable climate. Every report and every government pronouncement all say the same thing: the green transition will produce millions of jobs.

Yet it's not turning out that way.

It should be obvious that there are not as many green jobs as were promised, and, like in the rest of the economy, far more badly paid and 'bullshit' jobs are being created than well-paid, secure ones. To do otherwise the transition would have to push against the broader trend towards continued deindustrialisation and servicisation characterised by stagnant real wages and deteriorating conditions. The transition isn't counteracting these broader trends, however – it is leaning into them.

British Steel isn't suffering due to a lack of potential, but due to international competition. And there is no technological fix. Perhaps that is not surprising, given the dominance of China as the sole global green manufacturing superpower, one that also dominates steel production. China is certainly being set up as a threat to Western powers, producing too much and in turn consuming too little. While it could be argued that China is heavily subsidising its industries (as do many other countries), the positioning of China as an economic threat signals a deepening turn towards racist nationalism as a ruling-class project in the face of the climate crisis.

China does dominate almost all the major green industries. It manufactures almost 80 per cent of all electric vehicles and holds 80 per cent of the global market share for solar panels, as well as 66 per cent of the market share for wind turbines. China accounts for a staggering 35 per cent of the total gross world manufacturing production.[1] It is the largest producer of everything from cotton to cars to steel. One of the main reasons for the plight of both British Steel and Tata's Port Talbot plant is that China makes 54 per cent of the world's steel at a much lower cost – it's up to twenty times cheaper to produce steel in China than in the UK.[2] The idea that there ever was any hope for the British industry seems farcical.

But the idea that China is 'winning' the struggle for industrial dominance is a far too simplistic narrative. Like much of the rest of the world, China too is suffering from the effects of the transition and broader global manufacturing overcapacity. While older, heavier industries in China are already in crisis and whole regions deindustrialising, greener industries are already in a cycle of consolidation. China's fate is likely to be an echo of the West's, as the economy shifts towards a service model and growth winds down.

We can already see the shape of the not-so-bright new green economy. Broken promises and lost communities rub up against an escalation of the gig economy. Low-skilled workers and those living in rural areas are left behind. Contract work proliferates in the Global North, while few manufacturing jobs are created. Instead they stay in Asia, where working conditions remain poor, and workers must fight just to be paid. The labour of care will intensify, as will most women's double burden, at a time when the material grounds of the misogynistic backlash will only increase. Labour's share of global income will continue to decline. The transition will not do the political work of transforming the economy or securing real material justice for us. There is no technology that can save us.

These broken promises of green jobs are shaping the political and social landscape, further hollowing out the legitimacy of existing political systems and parties. They merge with the impacts of automation, rising precarity and the disintegration of the middle class, manifesting as a profound failure of the very idea of the future. This broader crisis of political legitimacy weakens the ability of unions to recruit and mobilise workers – it is as much a problem for the Left as it is for mainstream political parties. When confronted with this recalcitrant public void, politicians have largely responded with a right-wing populist agenda, seeking to mobilise people's fears around 'external threats', now increasingly including climate change policies themselves.

Broken Promises

As space for the future is squeezed by the impacts of climate change, the chorus of broken promises form the emotional and political basis on which the transition is taking place. Nowhere is this clearer than in the economic geography of the oil and gas industry.

On the north-east coast of Scotland, around the edge of the North Sea, workers in the oil and gas industry were told that as fossil industries wound down they'd find work in the renewable energy sector. Just over 27,000 people are directly employed in oil and gas in the North Sea, with the industry supporting a further 95,000 jobs across the region. This number is projected to halve by 2030. Endless NGO reports and government statements hail a green jobs boom for the region, with thousands of wind turbine jobs supposedly in the offing. Only one tenth of the promised wind energy jobs have materialised, however.

Currently there are around 38,000 people directly employed in the wind industry. And while reports suggest the sector could grow to employ nearly 120,000 people by 2030, this is conditional on 'dramatic support' from government, something that has been lacking in recent years. It certainly isn't coming from oil and gas companies themselves, most of which have invested nothing at all in the renewable energy industry. And they are unlikely to do so on their own, as profit margins for the industry are too low to prompt investment.

It is the same story for the solar industry. Manufacturers in Europe are cutting jobs, citing Chinese competition, while politicians respond with talk of tariffs and trade investigations. At the same time the picture in China is also one of consolidation – a euphemism for job losses – as overcapacity and intense competition provoke 'cost-cutting' measures.[3] The problem is that while there are not enough solar panels for the climate, there are too many for the market. Profit margins in Chinese solar firms have collapsed; nowhere in the supply chain is anyone making any money.[4]

There are similar fears for another supposed engine of green job growth: electric cars. Eighty thousand jobs in the automotive industry have already been lost in Europe because of the shift to electric vehicles. Forty per cent of the remaining jobs in the German sector are expected to go, part of the loss of a predicted 300,000 across Europe. Thousands of similar jobs are being lost in the UK and USA. And where jobs are not lost, the threat of redundancy is already leading to increased job insecurity, lower wages and worsening conditions. Beyond the factory gate, thousands more jobs are at risk along the complex supply chains and in supporting industries, such as auto mechanics and petrol stations.

There are two key reasons for this. The first is that electric vehicles just don't need as many workers across the supply chain. They have far fewer parts, and rely on batteries, not engines, meaning they require around 40 per cent less labour to build. Electric vehicle factories also make use of much greater levels of automation and robotics than existing car plants. The second reason is that as governments invest billions in national car manufacturing industries, more productive capacity is being created than can be profitably put to use, leading to declining profits and increasing numbers of mergers. The engine of capitalist development in the twentieth century, one that still employs 7 per cent of the European workforce, or 14 million people, is unlikely to be saved by electric cars.

Manufacturing has long existed as a crucial political horizon, holding out the promise of secure well-paying jobs and economic growth. This promise, made largely to men, holds specific importance following the waves of deindustrialisation over the neoliberal period, and the brutal austerity after the 2008 financial crisis. More than a promise of work, it sets out a future waiting on the other side of stagnation and economic decline. It promises revival.

Instead, for all the talk of a green manufacturing revival, US manufacturing will lose, at best, just over 100,000 jobs by 2030. Britain and Europe's manufacturing sectors are expected to continue to shrink while becoming more geographically concentrated. Where, then, are the green jobs we were promised?

Broken promises are worse than no promises at all. In place of hope for the future, what we find is a growing sense of betrayal. This betrayal risks becoming the dominant political emotion of the green transition, one all too easily mobilised by an emboldened far right. In the absence of the green future promised by a manufacturing revival, bitterness and resentment will grow as increasing numbers of workers and communities are abandoned.

The Installation Economy and the Endless Horizon of Low-Waged 'Green' Work

So far, what the green transition looks like is an installation economy.

Installation work is the labour of setting up and assembling systems and equipment on-site or in a specific location. From IT to construction to manufacturing, an underappreciated amount of what we think of as manufacturing or industry consists of setting up things made elsewhere. Within green capitalism, the labour of assemblage has displaced manufacturing as the key sector of job and business growth. A perfect example of this is the solar industry.

Employment in the European and US solar industry is booming. Most of this solar 'industry', around 80 per cent in Europe and the US, is a service and not a manufacturing industry. There are currently over 600,000 solar installers across Europe, with that figure set to climb to over 1 million by 2025. In the US, installation work is rapidly expanding, growing by 250 per cent over the past decade. And while there are some large employers, the bulk of the industry is based on agency work.

Solar installation in the US is like any other kind of agency work. Payments are often late, and workers frequently face days without enough money for bills or food, as well as rising debt and overdraft fees. Installation work involves moving from site to site, producing a large itinerant workforce, shorn of the protection of a local community. Workers report sharing hotel rooms or sleeping in tents and cars near remote sites, and many have horror stories of travelling to remote sites just to be told there is no work for them. Sick pay, days off, holiday pay – all of these are either non-existent or poorly enforced among many installation agency companies.

From glaziers and insulation installers to heat pump technicians, the green economy is an installation economy. Despite being labour-intensive work, it is better understood as part of the broad service sector rather than the manufacturing sector.

The service sector is far from uniform or homogeneous. It encompasses everything from law firms to selling burgers. As a category it often obscures more than it reveals. However, there are two unifying tendencies. The first is that, despite the variety of jobs and labour processes within the sector, there is a broad trend towards industrialisation – that is, subjecting workers to tighter managerial control and scripted performance. While this is clearly the norm for much low-waged and precarious work in the retail and hospitality sectors, automation is increasingly enabling the industrialisation of higher-wage professions. Often in the latter instance this manifests as a proliferation of bureaucratic procedures and processes that perversely hamper productivity rather than enhance it.

The second tendency is that the sector is broadly marked by lower productivity growth than manufacturing. At a fundamental level this is because services by their nature are more labour-intensive and interpersonal. And while automation and AI might succeed in industrialising services that have so far resisted other efforts at standardisation and routinisation, it is likely that the gains will be limited.

While boosters claim AI will make service workers more productive, it's more probable that it will continue to underdeliver, as have previous rounds of automation in services, leading not to productivity gains but more fragmented and deskilled work, contributing to rising precarity and stagnant wages. AI is more likely to contribute to a deskilling of installation work than to increase work productivity, substituting for labour instead of enhancing it.

Importantly, the installation economy differs from other service sectors in terms of how demand is generated. While some demand is driven by consumer choice in a 'pure' sense (i.e., compelled by the desire to reduce emissions or have some kind of energy autonomy), most demand is currently generated through government subsidies and legislation. As with other service industries, sector growth is predicated on the continued expansion of employment – it needs job growth generated in other sectors in order to stimulate demand. As an industry with high up-front costs, however, it also relies on there being sufficient demand from those wealthy enough to afford the installations. In recognition of high up-front costs, government subsidises purchases, making sector growth dependent on government programmes and legislation.

The installation economy, like other aspects of the green transition, is increasingly organised through political conflict under the guise of 'culture wars'. Government policy in Europe, Britain and the US at different times has driven uptake of solar panels, heat pumps and electric cars, while at others has dampened demand and frightened off investment, leading directly to job losses.

While much installation work is precarious and low-waged, this is not a universal labour condition. There are substantive differences between labour regimes, and within sub-sectors. While heat pump installers in Britain are paid well, as their labour is considered plumbing work and paid accordingly, the same cannot be said for solar panel installers. In the US there are substantive differences between unionised solar installation work on large arrays, and the far more precarious work of domestic panel installation. German workers in the installation economy are often paid slightly more than other skilled manual labourers, UK workers receive around the same, while US workers are often paid significantly less.

One crucial cleavage in the installation economy is between those workers who command a high level of autonomy thanks to their skills and qualifications, and those who labour under much more precarious conditions doing semi-skilled or unskilled work. This has an impact on wages and conditions both between industries and within them. Installation companies typically employ a mix of both types of worker, assigning them very different conditions and rates of pay (something long fought over within the construction industry). Such a divide can impede the production of a shared class perspective as interests diverge. This is all the more complex when work has been fractured into a series of small businesses, making workers into managers and bosses.

Many employers in the installation economy are small or medium-sized firms, often in strict competition with similar firms. This limits the upper ceiling of pay claims and reduces the bargaining power of employees, while introducing barriers to union organisation. At the same time work is dispersed around a region or country, making organising hard and limiting labour mobility between firms. And, as with other workforces in the Global North, the work is increasingly isolated and located within urban hinterlands and exurbs.

Installation work is often distributed through piece work or individual contracts, making it largely an extension of some of the least secure forms of existing service industry practices, not a break with them. In contrast to the promises of secure manufacturing jobs, the installation economy actively builds on the global neoliberal trend towards 'non-standard' work forms – zero-hour contracts, precarious and sub-contracted employment, informal or grey market jobs. Close to 60 per cent of all new jobs in the OECD since the 1980s have been non-standard, with the average level of such work across OECD countries now running at 30 per cent. The installation economy is unlikely to arrest this trend; if anything, it will accelerate it.

In place of promised manufacturing jobs, the transition economy creates installation work. Precarious, semi-skilled and largely controlled by small businesses, this work is unlikely to revive shattered communities or provide a significant number of local supply chain jobs. Instead, installation work deepens the trend towards an increasingly fragmented working class, and the total dominance of the service sector in the Global North.

Small Businesses, Landlords and the Suffering Agency of the Installation Economy

Although big firms dominated the headlines, the reality is that most green jobs are in small or medium sized firms. Around 60 per cent of employees in Britain work for small or medium-sized enterprises (SMEs), while the figure in the US is just under 50 per cent of all private sector jobs, and 68 per cent in Europe. Across all three, SMEs make up over 90 per cent of all businesses, making it the single largest sector. While SMEs are highly diverse, in general their wages are lower and their conditions worse than in larger firms; job satisfaction too is often lower. There is little reason to think that green SMEs are run any differently to the rest of the sector.

SMEs occupy a paradoxical position in the transition. On the one hand, the transition economy favours them. Installation work is an economic activity yet to be monopolised by big capital, and installation work is site-based work, enabling small companies to flourish in local areas. On the other hand, SME owners are already facing increasing costs as a part of the transition. They also often operate on thin margins and lack the scale or capital to make low-carbon adjustments to their practices. Paradoxically, the transition is both an opportunity and a threat to the sector.

It comes as no surprise that it's this class of business owners that forms the base of contemporary anti-transition movements. From attacks on traffic cameras to deranged conspiracy theories about 'climate lockdowns', the social base of such movements is the small business community.

It is tempting to see this community as a narrow demographic. And yet there are millions of SME owners – not including the millions of people who are self-employed and share similar economic relations. The small business community is significant, and includes two of the most politically active 'groups' opposed to the transition – farmers and landlords.

Farmers and landlords have been at the forefront of the recent backlash against environmental regulations and government policies in Europe. Farmers, especially small-scale ones, operate on razor-thin margins due to competition, rising costs and the power of large buyers such as supermarkets. Most European farms cover less than 100 hectares, with two-thirds under five hectares. Similarly, most US farms are small.[5] To stay afloat, farmers are driven into debt by investing in technology to achieve economies of scale, only for increased productivity to lower farm gate prices due to supermarket pressure. This precarious situation has left many reliant on government subsidies and struggling to make ends meet.

Landlords, though often perceived as wealthy, are predominantly small-scale operators. There are 9.72 million landlords in the United States, owning an average of less than two properties each. In Britain, over 2.8 million landlords own rental properties, with 43 per cent owning just one and most owning fewer than five. Many people become landlords accidentally, through inheritance or divorce, and supplement modest incomes with rental earnings. These 'petty landlords', averaging £24,000 annual income before counting their rental income, rely on rental income primarily for retirement funds or to bolster limited finances. In Germany, where 60 per cent of rentals are owned by private landlords, the story is similar. While in the US most land-lords earn around – or more – than average, a significant minority, largely comprised of 'mom and pop' landlords, earn below the average income.

These economic pressures have fuelled political turbulence. In 2023, farmers across Europe staged militant protests against environmental regulations, fearing they would be forced out of business. Their actions were linked to a surge of support for right-wing political parties, such as the rise of the neo-fascist Alternative for Germany (AfD), which capitalised on opposition to environmental policies like heat pump installations and gas boiler bans. Landlords, too, mobilised against such regulations, delaying policy implementation in Germany and Britain. In Britain, landlord opposition, enabled by their political connections, persuaded the government to postpone banning new gas boiler installations, showcasing how economic grievances are being harnessed by right-wing movements to gain influence.

One reason to think about farmers and landlords is to understand why they have largely set themselves against the transition. In an obvious sense, it's a question of profit margins – insulation, glazing and heat pumps all eat into profits, as do environmental regulations. But the main reason to consider them is that both the housing and agricultural sectors are flashpoints in the transition. Both are responsible for significant amounts of carbon emissions, and both have little to gain economically, in the short term, from the transition. Yet as with the broader SME sector, the transition will be made through them. SMEs are the key terrain of the transition economy buildout, giving small business owners, including farmers and landlords, incredible political power.

Small bosses often work within their businesses. In other words, they tend to think of *themselves* as the business. This is often terrible for workers, but it is not much better for bosses. They tend to regard their work not as self-exploitation, but as a kind of precarious autonomy. They experience work as an endless struggle to stay afloat. In the US and Britain, 20 per cent of small businesses go bust in the first year. Forty per cent of American small businesses go bust in the first three years, while close to 60 per cent fail within Britain. The self-employed are twice as likely to be living in poverty as full-time employees. This struggle is the reason why over half of SME owners report having poor mental health over the past year.

There is no security to be found in this class. Instead, work involves relentless vigilance. You need to be vigilant over your reputation and keep a permanent eye on changes in market conditions, on red tape and tax codes, on your savings and stock, on your suppliers, on your staff if you have them, and ultimately on your income as a precarious revenue stream. You are always at risk of failure.

As with the squeeze, this is lived as a steady crushing of possibility. It might be your business, you might control your work, but in the end there's really no choice: you have to pull long hours, you have to cut back at home to buy the new equipment or make the investment you need. There is never any way to avoid thinking of how the business is performing, or of the cost of every setback. What work looks like from here is a kind of suffering agency where the desire for freedom-in-work, for autonomy, manifests as a decision that comes with no realistic expectation of reward. For most people in this class, the desire for autonomy goes nowhere except to anxiety and paranoia – a suffering form of agency that is all too easily weaponised by a growing far right.

For the millions of people running these businesses, the transition isn't an opportunity, but one more thing to be endured and resented.

Going Round in Circles

While installation work is what the transition economy looks like in the Global North, it does not encompass the entirety of green capitalism as either a distinct sector or as a general tendency. One often overlooked aspect of the transition economy is the increasing adoption of 'circular' production models and principles.

In its broadest definition, the circular economy is a model of production and service delivery that aims to keep products and materials in circulation through processes like maintenance, reuse, refurbishment, remanufacture and recycling for as long as possible. It is generally used as a business model, although it is also attracting interest as a framework for reorganising government services and social economy enterprises. Despite there being no single definition of the concept, almost all its iterations operate on the basis of three principles: eliminating waste and pollution, circulating products and materials for as long as possible, and the regeneration of nature.

Essentially, circular production does two things. It restructures production to not only use less energy and fewer resources, but to design products that can be maintained or refurbished in order to last longer. And it pushes a business model that prioritizes maintaining control over products for longer.

Circular approaches often build on the principles of 'Lean' management and design in order to reduce waste and increase efficiency. They may also integrate services into the production model – repair and refurbishment are two services typically found in circular business models. Some approaches go further and look to turn products into services, in order to retain product value for longer (subscription models and leasing are common circular practices). An example would be a carpet company that leases carpets rather than selling them, or an appliance manufacturer who leases washing machines.

Europe and Britain have circular economy directives and plan to move as much of their economies as possible towards circular economic models. The US has been slower to adopt such frameworks, though many US-based businesses such as Dell Electronics have begun to modify their operations in line with circular principles. ING surveys indicate that 16 per cent of US firms have adopted a circular economy framework, with a further 62 per cent planning to do so in the near future. Globally, 26 per cent of businesses have adopted some form of circular production, with Europe and Japan at the forefront of this development.

It may be the case that the widespread adoption of circular production does realise significant carbon emission cuts. European proponents claim it could reduce material usage by a third and halve Europe's carbon emissions. It might also be the case, however, that efficiency gains are lost through economic growth and the expansion of production. Regardless, in practice the circular economy has a limited impact on job creation, despite promises that the green economy would generate bountiful good employment opportunities. What it does do is accelerate the move away from manufacturing towards services, and boost employment in waste management.

The waste sector is one of the 'big' green industries. Reducing the impact and emissions of the economy requires effectively managing waste streams. Intervening into waste streams enables a more thorough approach to materials capture – recycling – than reducing the final amount taken to incinerators or landfill (landfill sites produce 11 per cent of the world's annual methane emissions).[6] The waste sector is also critical to realising circular production methods, since capturing materials before they are discarded in order to keep them in circulation lies at the very heart of the project.

Waste is also a big employer. In Britain there are 120,000 waste workers, in the US there are 480,000, and 1.6 million in Europe. Globally, 6.7 million people are employed in the sector; the actual figure is much higher, though, as many workers are only informally employed in waste management, particularly in the Global South.

While wages and conditions can vary widely, most jobs within the sector are paid well below average rates, and conditions are poor. Workers have to handle toxic and dangerous materials, and accident rates are high. Hours tend to be long, and both precarious and temporary work contracts are common.

Between 50 and 60 per cent of employees work in waste collection. Collection is the lowest-paid role in the sector, often well below the national average – sometimes half as much. It is also a sector with high rates of migrant labour – between 15 and 25 per cent in Europe – where workers have low levels of tertiary education, meaning they often lack other employment options. The sector in Europe is largely public, though private companies are increasingly playing an important role, and are expected to take a greater market share through circular initiatives. With privatisation will come further work intensification, worsening conditions and anti-union practices – as is already common in the US, where two-thirds of the sector is managed by private companies. Finally, it is a sector that is exposed to the worst job-destroying aspects of automation. Estimates put the number of waste jobs already lost to automation in Europe at between 10 and 20 per cent.

In the UK, waste processing employs 20 per cent of all 'green' employees. The figure across Europe is 18 per cent. With higher recycling targets and a broader shift to the 'circular economy', this figure is predicted to grow, and temporary, low-waged, dangerous contracts are set to proliferate. Not exactly the sort of green industry politicians make stirring speeches about when selling their transition policies to the public.

On the Factory Floor

The installation economy defines the transition in the Global North. Yet conditions in 'green' manufacturing across the rest of the world are no better. If anything, the green transition is amplifying and accelerating negative trends already underway.

Worldwide, despite a surge in investment, industrialisation has peaked as a proportion of both global and almost all national labour forces, with the possible exception of China. While many explanations for this 'premature deindustrialisation' at a global level have been offered, by and large it is a problem of increased overcapacity, stalling demand, and the impacts of automation.

A third of global manufacturing takes place in China – double the US's share and three times that of Europe. Up to a third of all manufacturing jobs are in China, a figure predicted to climb to 43 per cent by 2050.[7] China hosts 80 per cent of the world's solar capacity, and manufactures two thirds of the world's electric vehicles, wind turbines and lithium-ion batteries. Much of this will be installed in China, too – it is expected that a full 60 per cent of all renewable energy installation between now and 2030 will happen in China, meaning there is more than enough demand to secure the health of the national sector.[8] At the same time, the low cost of Chinese products has led to a surge in installations in the Global South, at a rate well above those of Global North countries.[9] Scale, cost and quality have secured Chinese firms' market share, making them largely unassailable, even with the high tariffs being imposed across the Global North.

As China grows its share of global green manufacturing, the broader manufacturing sector confronts challenges akin to those in the Global North, including the hollowing out of former industrial hubs. Newer factories rely heavily on automation, leading to fewer manufacturing jobs. This shift exacerbates job precarity and undermines efforts to secure stable industrial employment. Meanwhile conditions in Chinese factories remain poor, with low wages, inadequate social protections, excessive overtime and unsafe working environments. Like many advanced economies, China also suffers from a number of other labour market issues, including an ageing workforce, a significant skills gap and the rise of a precarious platform economy.

While European steel and automotive industries blame China for declining market shares and job losses, the Chinese automotive sector is facing similar destructive dynamics. China produces a third of the world's cars and dominates electric vehicle production. Yet while sales of electric vehicles are strong, and there is a concerted effort to secure international market share as production outpaces local demand, the Chinese EV industry, much like the traditional ICE industry, is shedding jobs and closing factories. Strong competition has provoked a wave of consolidation, as companies struggle to secure market share and sustain profits. While sales reach new heights, workers protest over unpaid overtime, or the lack of severance pay as many smaller manufacturers go out of business. The remaining workers complain of increased workloads and dangerous conditions. In a portent of manufacturing work in a hotter world, SAIC-Volkswagen decided to turn off factory air conditioning to save on electricity. According to the workers, it isn't even turned on when temperatures top 34°C.

China also employs half the world's renewable energy workforce. Long hours and overtime are very much the norm, often because there is a lack of available skilled workers, perversely creating overwork and underemployment at the same time. In addition to the low wages and poor conditions, workers are often exposed to high levels of toxic materials such as sulphur dioxide, arsenic and cadmium during the production process. Similar toxic conditions are also prevalent at the other end of the solar panel's lifecycle, when it is stripped of useful materials and recycled.

This is not just hard and dangerous work – it is frequently coerced. Multiple reports have documented how much of the Chinese solar industry relies on forced labour, numbering in the millions, often in the Uyghur region. While the full extent of forced labour in solar production is unknown, it is plainly a very significant component of renewable energy production. Of course, forced labour is not restricted to China – the use of prison labour in the US is equally notorious. Prison labour has been used in the manufacture of US solar panels in the past, and rural areas often rely on prison labour to perform another crucial green industry task – firefighting.

Beyond the factory, work in the renewables industry encompasses a wide range of up- and downstream tasks, from mining to waste disposal and recycling. Precarious working conditions are prevalent across these green value chains, from biofuel production to solar and wind energy manufacturing, lithium production and waste disposal. Workers are exposed twice over to harm, experiencing not only the extreme heat and the toxicity of green industrial production and waste disposal but also the environmental damage resulting from extraction and material processing.

While in the Global North the green transition is touted as a means of creating clean, green manufacturing jobs, the reality in the actual manufacturing hubs of China, Indonesia, and beyond is that green industry is very much like its old dirty counterpart, and possibly worse.

Who Cares?

Most manufacturing workers, solar installers and waste collectors are men. The two demographics most in demand for green jobs are STEM graduates and manual labourers. In the UK, just 16 per cent of engineers are women, while women make up just 27 per cent and 26 per cent of manufacturing and energy workers. Men occupy the vast majority of 'green' jobs, from offshore wind industry jobs to climate finance roles.

All the sectors with government support and investment pouring in are predominantly male. Most of the predicted job growth – even if it's more fantasy than reality – is in male-dominated sectors. What we find in discussions of the transition is a political emphasis on what men stand to gain or lose, making the legitimacy of the green transition subject to broader misogynistic questions of the 'decline of men'. The green transition can look a lot like men talking and worrying about other men. As long as the transition remains politically grounded in the discourse of what's in it for men, this dynamic will only deepen. Given the material reality of broken promises, what looms within the transition is a par-anoid turn to misogynistic politics, deepening an already existing tendency within the Global North.

The political emphasis on manufacturing is disorientat-ing not only because transition economies in the Global North are better understood as installation economies, but because they are, more broadly, service economies. And this is not expected to change. Currently, service industries focused on social reproduction, such as health, care and edu-cation, are growing rapidly in most Global North countries. In part this is a demographic issue. Most countries in the Global North are ageing, and there are a number of inter-secting health crises, including mental health crises, that call for more care and health workers, while the transition itself is generating increased demand for tertiary-educated gradu-ates and training programmes.

Climate change is also intensifying the work of care. Confronted with the squeeze, the gendered labour of main-taining and feeding families becomes harder, more stressful. The physical and mental health crises produced by the impacts of climate change require more health care and mental health services. The transition economy has its own impacts, from demands for education to services for deindustrialised communities and newly unemployed workers. And as the cost of living increases and services struggle, the double shift demanded of women increases.

We've already seen how the climate squeeze will continue to accentuate the gendered division of unwaged care work. The transition also necessitates an increase in waged care, health and education work. But unlike manufacturing and other heavy industries, there is little government support forthcoming for these sectors. In fact, all three have faced severe cuts and lack of investment across the neoliberal period. This tendency has intensified since the 2008 financial crisis, and is not addressed in any transition policies or plans. Green capitalism is no different in its eagerness to exploit the cheap labour of women. Clearly, to matter politically in the transition, you have to matter financially to capitalism.

Green industries require new training and retraining programmes, alongside the expansion of university programmes relating to research, development and management. The need for more workers in the education sector is particularly stark, as the sector already faces staff shortages in all the OECD countries. This is not just due to intense workloads; it also reflects the relative and absolute underpayment of educators. The situation in the care and health sectors is similar, with persistent staff shortages, low wages, long hours and difficult conditions. All three sectors also employ far more women than men: in Britain, 75 per cent of teachers and health care workers, and 60 per cent of carers, are women, with similar proportions in other OECD countries.

As with other sectors, job growth doesn't mean job improvement or wage rises. Increasing the number of jobs in these sectors without reforming existing conditions will merely make women work harder for less in the green transition.

The need for more care amidst the broken promises of green jobs and the installation economy could make the green transition a losing proposition for women. Already burdened with a double shift, more unpaid labour will be expected from them, while they continue to occupy the majority of poorly paid roles within the care industries. In sum, the need for care will increase along with the stress of providing it.

Heat Stress

In August 2022, Columbus, Ohio witnessed a momentous teachers' strike led by the Columbus Education Association, representing nearly 4,500 educators. The four-day action called for the installation of air conditioning in overheated classrooms and better school conditions, including smaller class sizes and broader educational provision. The strikers were victorious and won new contracts, including the air conditioning stipulation and annual pay rises.[10]

In April 2024, tens of thousands of schools across the Philippines closed due to extreme heat. India shut schools in several states, while Bangladesh closed all of its schools. In Britain, where rising temperatures are creating unworkable conditions, the National Education Union, the largest teaching union, is campaigning for a legal maximum working temperature, as no such limit currently exists. In a union survey, 85 per cent of teachers said their building overheats in summer, with a third saying that temperatures became extreme.[11]

People work and study best at temperatures between 16°C and 24°C. When temperatures exceed these limits, not only does productivity suffer, but workers can die. The first impacts are often on performance: above 24°C, both physical and mental labour become more difficult and workers are less productive. By the time temperatures reach 30°C, workers are only half as productive and workplace accidents rise by between 5 and 7 per cent. Above 38°C and accidents rise by 10 to 15 per cent. At any point above 25°C heat stress can become a very real danger. Dizziness, fainting and heat cramps can occur and the risk of heat stroke or collapse dramatically increases. Above 41°C, delirium or confusion can occur, and blood temperatures can prove fatal. All these factors are already leading to worker injury and death. The International Labour Organisation (ILO) estimates that each year there are close to 19,000 work-related deaths from excessive heat, while almost 23 million workers suffer occupational injuries.[12]

As with the 'excess deaths' caused by heatwaves and high temperatures, official figures often understate both fatalities and other negative health outcomes. In the US, while officially 'only' forty people die each year due to extreme heat, unofficial estimates put the actual figure much higher – between 600 and 2,000.[13] The injury toll is around 177,000 per year, making extreme heat one of the top three causes of workplace deaths and injuries.

The stories of these deaths are often depressingly ordinary. A twenty-eight-year-old farm worker dies from heat stroke in Florida; a construction worker in Texas overheats and dies, while another Texan worker, repairing powerlines, dies of heat exposure. A postal worker dies on his delivery round. In France, six people die harvesting grapes. An Italian contractor dies while building a distribution warehouse.[14] For many in the installation economy, heat is an everyday menace along with falls, danger from electrocution, storms and, in parts of the US, alligators and snakes. Often this is made worse by the lack of legal rules on health and safety or safe work environments.

What's more, against the backdrop of a right-wing anti-climate and anti-worker political resurgence, government agencies are as likely to act against workers as they are to defend their own interests. Florida governor Ron DeSantis passed legislation in 2024 banning any municipalities in the state from enacting laws to protect workers from the heat. The same amendments also weakened child labour protection laws. Across the US, business has been lobbying against heat protection laws in much the same way as they have been pushing back against other environmental regulations, from plastic and chemical pollution to nitrogen run-off, all contributing to the toxic dangers of work.

Soaring workplace temperatures is one of the most obvious ways in which climate change will reshape the conditions of work, making many industries fundamentally more dangerous and requiring, as teachers are demanding, structural and policy changes. These could include anything from the installation of AC to adjusting working hours in some industries (as in the Spanish agricultural sector) to abandoning certain kinds of work during specific hours or even seasons.

While all those working outdoors in agriculture and construction are at risk, the danger is even higher for migrant workers in those industries, because of existing rates of exploitation and low levels of protection. Extreme heat is one more in the long list of issues confronting migrant workers. The risk is also substantially higher for older workers, for two reasons. The first is that older people aren't able to sweat or cool down as efficiently as younger ones. The second is that many have medical issues – from diabetes to heart conditions – that are aggravated by heat stress. As workforces age in the Global North, there will be an increasing cost in lives, one that will in turn exacerbate the squeeze on care industries.

Mining and Security

There is no transition without the 'green' resources and the growth of the extraction business. The green transition will produce a huge surge in demand for metals and critical minerals. The global extraction of the necessary raw materials is predicted to rise by 60 per cent over the next forty years. Increases in specific metals range from doubling the amount of copper produced based on today's figures by 2050 through to a 900 per cent increase in lithium production by 2030. Demand for nickel, cobalt and rare earth elements, as well as aluminium, will rocket.

Digging, blasting, mining; hidden from view, it is the dirty work of the green economy. Globally, mining employs 19.3 million people and produced US$711 billion in revenue in 2022. The oil and gas industry – at its extractive frontline – employs 12 million people, and the entire industry has produced US$2.8 billion in profits every single day for the past fifty years. Quarry work and forestry employ another 37 million.

Working conditions and wages in the planetary mine vary widely, from highly paid in some countries in the Global

North to something akin to slave labour in many mines across the Global South. Yet while there are significant mining and extractive projects across the Global North, the majority are in the Global South, not only reinforcing colonial relations and uneven exchange, but guaranteeing that mine work is, for the most part, poorly paid and takes place under harsh conditions.

It is also changing. Where economical, mining work is becoming more automated, so that even poorly paid jobs are disappearing, robbing local communities of one of the few benefits of the mines. While the threat of job losses in mining is so far concentrated in the Global North, where it's estimated that up to 45 per cent of existing jobs are on the line, this does not mean the Global South will be spared. The world's first fully automated mine is in Mali, and Chilean copper mines already have autonomous haul trucks.

Those lucky enough to be exploited face other threats. Like many industries, mining will be exposed to the direct impacts of increasing temperatures, driving up costs and risking increased accidents and death for workers. Water supplies to mines – often crucial for their functioning – will shrink, putting operations at risk. And as many are remote, flooding and other extreme weather events pose particularly high levels of risk for workers.

Mining, like other extractive primary industries, doesn't take place in a social or ecological vacuum. And much of the mining necessary for the transition will involve breaking new ground. Both existing mines and future mines will face stiff opposition – for each project invades a community or habitat, prompting resistance from local communities and environmentalists, if not whole regions, peoples and nations.

In 2023, mass protests rocked transition mineral mines in Panama, Peru, Argentina and Serbia. Protesters declared victory in Panama, while ongoing demonstrations in Serbia have put a proposed lithium mine into question. In South America's lithium triangle, riots and blockades are commonplace as local communities resist the despoiliation of their towns and farms and the poisoning of their water supplies. The Environmental Atlas project shows over 300 current anti-mining protests across Latin America alone.[15]

Opposition to mining isn't an aberration or exception. Protest and local opposition is an expected feature of extraction as a process. Any proposed project or investment must consider the cost of resistance and measure the likelihood that the project will fail. From pipelines to mines, both business and government figure in opposition. Should that arithmetic suggest that resistance will be strong, or that agitation could stop the project, then business won't invest, and governments will often not move ahead with the plans.

The labour of working the planetary mine, then, is not just excavating and moving the earth, but overcoming resistance to mining.[16] Someone needs to enforce the clearance of village communities and guard the site. The extractive frontier is often a war zone – and this war zone is someone's workplace. As the new enclosures of the neoliberal era proceeded, they were reinforced by a sharp increase in military and security spending that has often gone unnoticed. Global military spending went from US$600 billion in 1980 to US$2.2 trillion in 2022. Alongside this boom in military spending, the private security industry exploded, guarding everything from mines and warehouses to farms and water reservoirs. And it is worth at least US$235 billion per year – a figure that is set to almost double by 2030.[17]

There are upwards of 25 million private security guards globally, not including guards working under informal contracts or on the edges of legality. Globally there are around 11 million police officers, many of whom often undertake mine security, especially the repression of protest. In many countries the military work to protect and secure mining interests, meaning a not insubstantial part of the 27 million personnel within the global military workforce are also implicated in defending extractive work.

Border security, for its part, is worth US$49 billion per year and globally employs millions of workers. Posting 7 to 8 per cent growth rates per annum, the industry has a bright future if UN projections of climate refugee numbers are to be banked on. Conditions across this type of security labour vary widely, from profoundly poorly paid – guards for the Qatar World Cup received less than four dollars for a twelve-hour shift – to incredibly lucrative: Blackwater mercenaries earn as much as US$250,000 per year as a starting salary.

When we factor in the intense militarisation of the police, making them into an occupying 'blue army' that imports well-drilled anti-insurgency tactics from the Global South to the centres of the metropole, it's clear that the broader transformation of the global economy has relied on a historic wave of militarised repression. We could go so far as to say that, rather than financialisation, the neoliberal period was characterised by militarised accumulation, or accumulation by repression.[18] As a condition of the green transition, one that enables social security in the Global North and a steady supply of raw materials for green industries, it is also an industry set to grow and a form of accumulation set to intensify. Work as a border guard, mercenary, or 'security operative': those are the green jobs they didn't tell you about.

The Net Zero Economy

The new green work looks very much like the last wave of job creation – Amazon warehouses, Uber contracts, the light industrial assembly work that proliferated in the urban hinterlands, or the labour of semi-trained contractors maintaining cable networks and internet infrastructure. Or, as in the case of the manufacturing of green technologies, we find a continuation of existing forms of labour exploitation in the free trade zones and mine sites of the Global South. Across both, the green jobs boom is not enabling a renaissance of labour organising but rather a deepening of neoliberal trends.

Unlike previous waves of job creation, however, a significant amount of the new roles will not be long-lasting or permanent. As the buildout of renewable energy infrastructure proceeds, it will inevitably peak. At that point, maintenance and (eventual) replacement installation will constitute the bulk of the available work. To take one example, while there may be a need for 1 million solar installers between now and 2030 in the US, far fewer will be required after that.

To be sure, there is also a surge in new professional and scientific roles, but nowhere near enough to soak up university graduate numbers. And many of these roles are just 'bullshit' managerial and compliance roles – make-work, filling in government forms, and adding managerial greenspeak to internal documents and corporate reports.[19] From sustainable fashion consultants to compliance officers, this work is not actually green work, and beyond paperwork and PR offers little to corporate bottom lines, let alone social value.

The gutting of remaining industrial centres in the West, coupled with the shift to an installation economy, implies a fall in indirect jobs. Compared to industries like automotive manufacturing and petrochemical companies, installation work tends to produce far fewer jobs in the supply chain or broader economy. And with China's continued dominance, many of the supply-chain jobs that are produced will emerge, and stay, there. Accordingly, the installation economy won't functionally revive the stagnant economies of Europe and will likely underperform in terms of job creation in the US.

This does not mean that China, and the economies of the Asia-Pacific region, will host a manufacturing boom, at least not one that might make up for the losses in heavier, carbon-intensive industries. We are already seeing the first stirrings of a green deindustrialisation in parts of China, with rust belts proliferating across the once booming north-east of the country, and where opioid addiction now proliferates in the place of manufacturing.

China is currently betting on maintaining its share of global manufacturing – even increasing it – so that, notwithstanding the move towards dark factories full of robots and the forward march of AI and automation, there should be large numbers of manufacturing jobs in the decades to come. This is far from certain, however, and relies on the continued expansion of China's economy. Across the rest of the world, it is very likely that the manufacturing workforce has peaked, even in lower-income countries.[20]

As we add up the broken promises, it becomes clear – there is no technology coming to save us.

3

The Transition Economy

If the transition economy feels somewhat familiar with its precarious contracts, growing service sector and persistent cost-of-living crisis, that's because it's built with salvaged parts. The transition economy is not a decisive break with neoliberal policies and political imperatives: it mixes up past tools with borrowed policies and desperate measures in a series of ad hoc policies, in response to the unique problems of waning US hegemony and a stagnant global economy.

Confusion over the transition comes from misunderstanding what it is. As I write this chapter, much ink is being spilt over the idea that somehow it's shocking that the world is on course for 3°C of global warming, that somehow scientists have misled us, and the transition has failed. Yet any sober assessment of the policies actually implemented by world governments has only ever come to the same conclusion: we are on track for 3°C even if governments stay the course.[1]

Campaigners and scientists have long aimed to set hard targets as a way of encouraging governments and businesses to act faster and more decisively. Yet all too often this has veered into uncritical cheerleading for the slightest hint of progress. Capitalism is built on irrational exuberance, yet commentators have often strayed beyond this into realms of outright fantasy, proclaiming the peak of emissions is near, and the future likely saved. We need to name this for what it is: a liberal utopianism that fervently believes that the state and the markets not only can but will act rationally when presented with decisive knowledge about an imminent threat. It's a belief in the idea that the powerful will 'do the sensible thing' when necessary.

Because this is not the case, the transition is called out as a false transition, or a lie. And the climate solutions that are proposed amount to little more than policy ideas for governments that will never be formed – rational governments, governments capable of doing what is necessary against political realism. Fantasy governments.

The transition economy is not a set of rational policies that aim to arrest climate change at 2°C, let alone 1.5°C – and we shouldn't judge it on that basis. It is instead two things. It is the recognition that enough powerful people in government and the business sector believe climate change is a real threat and, at the same time, it is the belief that this threat holds the promise of economic revival.

The idea that action on climate change might be the solution to a broader set of economic problems within the economies of the Global North dates back to the mid-1990s. The emergence of a policy-focused cohort of think tanks and NGOs in the shadow of neoliberalised social democratic parties led to the development of a novel green Keynesianism in contrast to the existing punitive government policies. There was a frenetic production of detailed policy documents, utopian in scope, setting out how environmental limits could be reconciled with job creation and other social policies.[2]

Aspects of the idea of a 'Green New Deal' moved from the margins to the mainstream after the 2008 financial crisis, where building new 'green' infrastructure was hailed as a solution to the Great Recession. From Deutsche Bank to President Obama's top economist Larry Summers, green Keynesianism became part of the broader economic policy debate. In Britain, a Green New Deal Group formed to campaign for the policy's adoption, while the economist Lord Stern authored a key review in 2006 of the economics of climate change for the British government, just prior to the government formally adopting legislation to reduce carbon emissions by 80 per cent by 2050. The report argued for a massive green Keynesian project as the best way to tackle climate change. Other iterations of the Green New Deal can be found in South Korea's 'green growth' strategy and Obama's 'cash-for-clunkers' programme. While some elements of green industrial policy would appear over the course of the decade after 2006, it was with the series of net-zero and climate policies that were adopted after the signing of the Paris Agreement in 2015 that the framework of the transition economy began to emerge. By 2019 the EU had adopted its Green Deal programme, and in 2022 the US passed the Inflation Reduction Act (IRA), based on the Democrats' 'Build Back Better' policy framework. Similar proposals and plans have since been adopted around the world. China had already implemented a much more robust set of green policies in its 2006 five-year plan, cementing this approach in the concept of 'ecological civilisation' in 2016.

This collection of policies doesn't form a coherent global plan. Nor could it, given the wide regulatory and ideological differences between the main economic actors. But there are common elements, particularly amongst the countries of the Global North.

The framework of the transition economy comprises a collection of government policy interventions that aim to bolster specific industries and direct national economic production through a mix of tax breaks, subsidies, support for research and development, protectionism and regulatory intervention, with the latter often aimed at transforming patterns of domestic consumption. As a solution to climate change, at its core lies the promise of 'electrifying everything', of using research (under the mantra of innovation) to 'fill the gaps' to enable everything to be electrified. As a solution to economic woes, it emphasises not only investment in 'frontier technologies' that promise improved productivity and economic growth, but a wholesale change in governance to guarantee profitability.

The transition aims to create an economic atmosphere conducive to investment in both capital expenditure and research leading to innovation. This is not the government directly investing in production; rather, the intention is to encourage business to invest. In the most immediate sense, the green transition amounts to the privatisation of climate change.[3]

While ostensibly in continuity with neoliberal policies, this privatisation takes place in an openly anti-market fashion. Governments give out market-distorting subsidies to favoured companies; tariffs and protectionism are once again the order of the day; monopolies and rentierism are not only tolerated, but actively encouraged. The market is

not the means by which prices are set – instead, governments increasingly intervene to set prices in order to enable
companies to profit.

At the same time, the transition economy names more
than an emerging paradigm. It is also a general tendency
within capitalism, and names a specific fraction of capital
that is directly involved in the production of green technologies and industries and that has its own interests and alliances.

An End to Growth?

It is now generally accepted that following the economic and
political crises of the 1970s, the economies of the Global
North have become largely stagnant.

Stagnation describes a period where there is minimal or
no growth in productive activity, often coupled with a
decline in profitability, persistent underinvestment, rising
unemployment and heightened inequality. There is little
need to rehearse the evidence here: economic growth rates
have been low compared to the post-World War Two period,
often 2 per cent per year or less; productivity growth has also
been low, and real wages have stagnated, while economic
inactivity and underemployment have both steadily
increased. Business more often than not prefers to speculate
than invest productively.

Stagnation went mainstream as a diagnosis after the fall-
out from the 2008 financial crisis. The crisis shattered the
illusions of economic dynamism in the Global North, leading many to declare that the economies of the Global North
had entered a period of 'secular stagnation'. What makes this
period of stagnation 'secular' is the fact that it is not an aberration or driven by external events, but is an expression of a

deeper, structural set of problems. Rather than setting out a full account of secular stagnation, in order to understand the transition we only need outline some of its essential features.

One of the key accounts of contemporary economic stagnation has been developed by the Marxist economist Robert Brenner.[4] In Brenner's analysis, the post–World War Two boom ended as the global economy became saturated with industrial manufacturing capacity. Where for a brief time the destruction wrought by World War Two enabled mutually beneficial national growth, by the 1960s there were already signs that international markets were becoming highly competitive, with key industries exhibiting more productive capacity than could be used profitably. The resulting glut drove down profit margins, leading to increasingly politicised trade wars, market consolidation, and ultimately a war on workers, all underpinned by government support, in order to restore profit shares for major producers and shareholders.

A number of factors were introduced to increase productivity and cut costs, from containerisation, enabling the global dispersal of the factory, to automation and new managerial approaches such as Just-in-Time production and Lean management. All of these generated increased productivity, but at the expense of jobs. Worryingly for proponents of automation and AI, it was largely this round of automation and job-destroying productivity increases that shifted the weight of employment from manufacturing to services in the Global North, and not the process of 'offshoring' jobs to the Global South.

The impact of these changes, combined with government efforts to weaken labour and increase corporate profit shares, was to drive down productivity growth and increase inequality, further depressing economic growth. Faced with slow to no growth, and weakened demand, business became increasingly reluctant to put money into productive investments, preferring financial speculation to the lower returns of manufacturing.

Giovanni Arrighi usefully expands on Brenner's account, bringing the question of politics back into the economic dynamics of stagnation. Arrighi argues that the social and economic conflicts within the Global North during the first half of the twentieth century drove up costs and hurt profit margins. Subsequent demands for higher wages and better conditions, coupled with demands for socio-economic inclusion by civil rights and feminist movements, led to a crisis of control for governments, and further reduced profit rates.[5] At the same time, the post-World War Two period was marked by surging anti-colonial movements, curtailing the free passage of (Western) capital and putting the global hegemony of the US at risk. Here too costs mounted as Western governments, led by the US, waged increasingly expensive wars on liberation movements across the Global South, all while the frontiers needed by capital became increasingly hard to access.

Overcapacity in this context is as political as it is economic. Denied the ability to export their way out of social and economic conflicts, and faced with increasing costs at home, the only possible outcome was a profound economic crisis. Secular stagnation, in this reading, continues, insofar as these conflicts remain unresolved at a global level.

A second feature is what we could call the Solow Paradox. Innovation since the 1970s, particularly in digital and communication technologies, has not given rise to sustained improvements in productivity. Productivity growth has remained low in most sectors, a fact that led the US economist Robert Solow at the end of the 1980s to state, 'You can see the computer age everywhere but in the productivity statistics.'[6] Robert Gordon built on this contention, arguing that there has been a structural decline in the pace of innovation. Gordon contends that most of the technological advances we will ever have in terms of productivity, speed and power already came in with the industrial revolution, meaning only marginal gains, made at great expense, are left. As economic growth is driven by productivity gains, a slowdown in the rate of innovation has led to a slowdown in economic growth.

While this seems paradoxical in an age of exploding AI research programmes, a renewed space race and unbridled veneration of tech-bro entrepreneurs, innovation that leads to greater productivity is increasingly rare. Indeed, most 'new tech' is focused not on innovation but on making incremental adjustments to existing products, generating ephemeral 'customer experiences',[7] or on locking customers into corporate 'ecosystems' in efforts to capture markets and reduce competitive pressures.

Research is also getting harder and more expensive, with more researchers required per patent each year. Worse, many digital technologies, including the much-vaunted AI, have already become 'enshittified'.[8] Cory Doctorow coined the word enshittification to describe a pattern of digital decay, whereby products such as online platforms decline in quality as service offerings become increasingly adapted not to users but to business customers, leading to services being eroded in the interests of profits.

Enshittification doesn't just impact online services. As software and subscription services come to dominate consumer purchases – a trend that's only set to rise through the circular economy and wholesale electrification – enshittification spreads to absolutely everything, from cars to clothes to food. And that includes the transition economy.

The third aspect of the slowdown in growth is people. The shift to a service sector–dominated economy reduces productivity growth, as services are less amenable to technologically driven gains.[9] Services also tend to produce fewer additional jobs than manufacturing. Additionally, there has been a slowdown in the growth of what is often called 'human capital' – that is, both the number of workers measured in labour market participation rates, and levels of education. Economist Dieter Vollrath contends that we have reached a plateau of 'human capital', and are at a point at which society has maximised the productivity gains achievable through improvements such as education, skills and training.[10]

More broadly, the Global North suffers from a profound 'crisis of care',[11] where the social resources necessary for the reproduction of physical and social life – namely, the unwaged labour overwhelmingly performed by women, as well as people's own physical and mental resilience – have been exhausted by government austerity programmes, work intensification, inequality and the relentless expansion of capital into every aspect of our lives.[12]

To this we can add a growing chorus lamenting the impending peak in the global population as yet another catastrophe in the making. People are living longer and having fewer children, meaning that many countries' populations are on average getting older. This process is profoundly uneven, with populations in some regions and countries already shrinking, while with others it is something yet to come. But the global population will certainly peak in the next decades, entailing a huge array of social, economic and health challenges for an economy built on the premise of ever-expanding human numbers. Dozens of countries will have shrinking populations by 2050, while the global population could peak as early as 2060, beginning to decline soon after.

The combination of the plateauing of human capital and an aging workforce means capital faces a looming labour power peak, as labour becomes both scarcer and more expensive. Given that one of the most critical inputs for capital is cheap labour, this signals a profound epochal crisis on the horizon.[13]

For us, however, it comes as a further turn of the screw, compounding the existing crisis of social reproduction in the Global North just as climate change devastates our health and our communities.

Innovating out of Stagnation

The transition economy is not a solution to all of these issues, although its most ardent supporters contend that it could be. For everyone else, the transition offers more modest promises. New products, coupled to government legislation, will overcome the problem of overcapacity without industry facing obliteration. Investment will return, as research in key new technologies overcomes the innovation decline. New technologies will bring productivity gains to services.

However, none of the policies or technologies that underpin the transition tackle the persistent crisis of social reproduction across the Global North, nor the issue of demographic change. The transition continues the tradition of 'trickle-down' economics, leaving inequality unaddressed. And it leans into the monopoly power of existing industrial and tech corporate behemoths.

It's worth noting the scale of the challenge. The International Energy Association (IEA) has estimated that nearly half of all future carbon emission reductions must come from technologies either in development now, or that are yet to be developed. That is, half of the technologies needed to ensure the planet remains habitable haven't been invented yet. Accordingly, the focus of most transition policies is on generating investment in these missing technologies. Governments tend to use a combination of four basic approaches.

The first is the use of subsidies, infrastructure spending, tax breaks and directly funded research programmes to encourage business investment in both new factories and services and in developing new research. Despite neoliberal rhetoric to the contrary, these measures have long been a staple of contemporary economic governance – agriculture being the most obvious example.

At times this looks like a kind of green protectionism for key industries, where government backs 'local' companies against international competition by spending big on supporting them. At other times, however, it looks less like protecting national industry and more like bribing international businesses to set up shop locally. Or, worse, bribing them so they don't close completely. This is the case in Britain for both the steel and car industries, where successive governments have spent millions trying to convince companies to keep some production in the country. In a world awash with overcapacity, companies want handouts more than they want protection.

The second is to actively create markets for green technologies. From bans on new petrol cars and gas boilers to mandating carbon emissions reductions and waste minimisation – these all create markets for transition goods and services. While often attacked by fossil fuel businesses, they do create an economic incentive to invest in future industries. Yet, as we saw in Chapter 3, these mechanisms are also often the most vulnerable to changes of political fortune, corporate lobbying or far-right anti-climate campaigns.

The third approach is price fixing, usually undertaken in two ways. The first is to use tariffs and other border mechanisms – or even trade embargoes and national security legislation – to ensure a minimum national price for various goods and services. Tariffs on electric cars from China, for example, are meant to ensure that US or European manufacturers can sell their cars at a reasonable profit, shielded from price competition. This tactic often leads to trade wars, not only heightening international conflict but, following Brenner's thesis, further increasing global overcapacity. The second way involves the widespread practice of guaranteeing prices for green products. For example, the low profit margins of most renewable energy schemes – on average just 5 to 8 per cent, compared to more than 15 per cent for oil and gas – have induced governments to guarantee higher prices per energy unit. That is to say, they have agreed on a certain level of income and profit for renewable energy generators. The reason being, of course, that businesses are interested in making money. Cheap energy doesn't interest them – a decent return does.[14]

As with other minimum price controls, these price floors support producers over consumers, including other business consumers of energy, meaning we all pay a higher price to ensure these sectors remain profitable. The broader impact of tariffs and price floors is to make the transition more expensive, and thus slow the entire, already inadequate, process down.

The fourth mechanism is to actively block opposition to 'new investment opportunities'. What this means in practice is tearing up environmental regulations, making exceptions to health and safety as well as labour laws, weakening planning rules and actively cracking down on protests to make investment safe for investors. After all, the transition economy is one that promises a boom in new factories, mines, power cables and windfarms. It is supposed to herald an explosion in building and construction – all of which must take place somewhere. The factory might be next to a river, or the power substation in a woodland. No matter where the development, there will likely be local community opposition as well as pushback from environmental and conservation groups.

Public opposition can make private investment in big infrastructure projects risky. Communities can organise against the imposition of projects, such as new roads or power transmission lines, in their local area. They can also demand that the companies managing the infrastructure do it properly, in which case shareholders receive fewer dividends and the company makes less profit. In both cases companies lose out. It doesn't matter whether opposition is reasonable, or if other options exist. Opposition worsens the investment environment – so by this logic it must be stopped.

Governments are already responding to the threat by removing the means by which people can legally protest and by pre-emptively waging an ideological war against so-called NIMBYs (Not In My BackYard). It doesn't matter that the original NIMBYs were fighting toxic industries and rapacious slumlords. What does matter is that opposition to business investment be made as difficult and minoritarian as possible.

By removing safeguards, these new projects and pieces of infrastructure will be built to the lowest possible standards to ensure the highest profit margins. This could invite more accidents and the generalised enshittification of the green economy, or it could mean the environmental and climate impacts of the transition buildout will be much worse. Attacks on local democracy and public concerns will deepen opposition to the transition, even among those inclined to support it. We are already seeing this take place in campaigns against wind and solar farms as well as new transmission infrastructure across Europe.

There is a temptation to see all four approaches as simply a matter of derisking investment, in line with the 'Wall Street Consensus' that has emerged around global development.[15] It is, after all, how governments expect to convince companies to create the transition for them. But this combination of approaches is far more than just a derisking programme. Similar to military spending, the transition economy amounts to the production of a vast captive market, formed by government targets and state-backed monopolies.

Much of this captive market will be dominated by monopoly conditions that encourage green rentierism. Rentierism describes an economic system where those that monopolise key assets like intellectual property, natural resources or digital platforms are able to extract economic rents. Rent refers to income derived from the ownership of a scarce resource or asset, rather than income from productive activity. A number of recent arguments have been made that Big Tech and the rise of platform capitalism have created a situation 'beyond capitalism', one verging on a feudal system, because they tend towards rentierism.[16] However, the shift to a rentier economy was already firmly established during neoliberalism, with infrastructure and contract rents becoming rampant. The transition will only further entrench this trend.

Ultimately, the hope embedded in government transition programmes is that these incentives will spur a surge in investment in new productive capacities, green industries and the application of novel technologies to existing industries including the service sector, to boost productivity and cut emissions. But there are serious doubts as to whether or not this will, in fact, lead to growth-creating innovations. It's also not clear, given the failures of many of the new technologies that have been declared essential to the transition such as carbon capture and storage and modular nuclear reactors, whether they will even deliver on carbon reduction goals.

Trade Wars and Green Dependency

Across the Global North, billions have already been invested into new companies and industries, as well as research programmes for the transition. By 2030 both the US and Europe expect to mobilise over US$2 trillion in total. While this is undoubtedly a large sum, it is dwarfed by the total required to establish a net-zero global economy by 2050. At time of writing, to hit that target the world would need to spend US$9.2 trillion every year until then.[17]

Billions are being invested by other countries as well. However, the real global leader of the transition economy is China, which invests half a trillion dollars per year in transition technologies and infrastructure. China is also increasingly investing in renewable energy projects across the Global South, financing nearly ninety in Africa alone since 2021.

There is no transition economy without China. The low price of most Chinese greentech has meant that the installation rate of renewable energy across the Global South is growing at over double the rate of the Global North. And it's doing so despite a lack of direct economic and financial support from traditional 'development' lenders such as the World Bank. China increasingly dominates not only green technologies but most major technologies from machine tools to commercial aircraft, all of which are essential to the prospects for economic growth. There is some debate over whether or not China is challenging the US's 'tech supremacy' across other frontier technologies, most notably AI, which dominates contemporary tech investment. While some recent reports suggest the US is lagging behind China in a wide range of critical technologies, others suggest such claims are overstated and part of a broader project of Chinese 'containment'.

Economic competition between nation-states and regional blocs has intensified since Donald Trump's first presidency in 2017.[18] Prior to this, the US had pursued a 'soft power' approach, one shaped by business interests in gaining access to Chinese markets. Both Trump and Biden levied numerous tariffs and trade-limiting interventions against Chinese goods and investments, in the context of rising economic nationalism across the Global North. Yet talk of deglobalisation and economic decoupling is dramatically overstated. While there has been a drop in direct trade between China and the US, this has not led to a breakdown in global trade.

Indeed, a significant amount of trade between the two countries was rerouted via third countries, while China expanded into new markets, leading to an overall increase in Chinese exports. At the same time, global trade has continued to expand, reaching record levels in 2024. China's recent economic slowdown has led to its share of global trade falling from 18 to 16 per cent, but this is still almost double the US's share.

While most of the barriers put up by the US so far have had a limited impact on China's overall trade – although we may see a stronger effort during Trump's second term – some, such as the restrictions on microprocessors, have had a powerful political impact. The US is attempting to deny China further economic growth in key areas, including sophisticated high-tech chips crucial for most complex digital systems, such as AI.[19] Bans on sales to and collaboration with China in 2022 have effectively locked China out of some aspects of the global computing economy.[20] Yet given the scale of Chinese research and investment, it won't be long before China is able to manufacture its own high-end chips.

Beyond the Global North and China, there has been much talk of countries of the Global South leapfrogging past the fossil-fuelled economies of the West into a clean green future. While the growth rates of renewables installation offer much hope here, as do the stirrings of a new non-aligned movement, the emerging reality is quite different.[21]

Green industry is by and large not relocating to the Global South, and those enterprises that do are often foreign-owned. Hopes for a technology transfer and the creation of national industries have largely been dashed. Green industry in the

Global South is, at best, a matter of assembling components made elsewhere. For the most part it, too, is an installation economy. This comes on top of the premature deindustrialisation of existing heavy industries, hit both by global overcapacity and trade barriers supposedly promoting a green transition.

What has grown are extractive industries, locking the Global South in place as a vast site of excavation for the Global North's transition economy, as a supplier of all the key raw materials, from food to water, and, increasingly, renewable energy.

When organisations such as the IEA declare that the extraction of the raw materials needed for the transition must expand by 60 per cent to meet demand over the next forty years, it's often unappreciated what this means in concrete terms. We are talking about billions of additional tons of materials, including iron, aluminium and copper as well as rare elements such as molybdenum and vanadium. This figure doesn't include the total volume of earth moved, which will be far, far higher – close to 350 billion tons of toxified mining waste.

This new round of extraction is expected to create an additional 400 mines, with upwards of 70 per cent of them located in countries of the Global South. But extraction is not limited to critical minerals. Food and animal feed accounts for one-third of the total value of global commodity exports, with many countries in the Global South significantly dependent on exports for their national income. The production of food and feed, as well as most mining processes, also use huge amounts of fresh water, effectively creating a virtual water market from the Global South to the North.

Traditionally, to this list of exports we would add fossil fuels – oil, gas and coal, all of which are still extracted from the Global South in massive volumes. Future energy exports won't be confined to fossil fuels, however. Billions of dollars are being invested in renewable energy projects across the Global South, not to tackle energy poverty but as export-orientated industries. Projects such as those established in Morocco by German and French companies to produce solar energy and green hydrogen for export, valued in the billions and with few local benefits, are being replicated across the Global South.

Finally, to this list we could add land itself, not only as a source of timber or food, but as a carbon sink. Land is crucial to most transition plans as a means of reducing the level of greenhouse gases in the atmosphere. The biosphere draws huge volumes of carbon from the atmosphere every year – although this process is starting to break down in a number of places. In order to reach net zero, many countries and companies rely on existing forests and other landscapes to offset their emissions. Given the scale of emissions in the Global North, many companies purchase carbon offsets from countries in the Global South, meaning they pay to use the forests and land of the South in their own carbon accounting, effectively blocking them from being used for any other purpose. That carbon offsets don't really work to reduce emissions is not the point here – the point is that the transition has enabled yet another means of extracting value from the Global South to ensure the continued overdevelopment of the North.

Far from initiating a round of global green development, the transition is not only reinforcing existing patterns of unequal ecological exchange – it is deepening them.

In an obvious sense, it is clear the countries of the Global South send far more resources to the countries of the North than they receive. If we include the damage done in the process of extracting those resources, the trade in ecological terms is profoundly one-directional. Adding the value of energy and labour time to that of raw materials and embedded water, we can see there is a net flow of not just resources but value and money from the Global South to the Global North. This unequal exchange has been calculated to be worth close to $2.2 trillion per year.[22] In the context of widespread economic stagnation, the value of these flows from South to North has been more important to the economies of the North than their own gains from economic growth.

The transition does not seriously challenge this dynamic. Rather, it is already locking nations across the Global South into a new arrangement of green dependency, where the typical relation is one where Southern countries exchange critical minerals for solar panels from China, then export the energy from vast, foreign-owned solar arrays to countries in the Global North, thus generating foreign capital in order to buy essential equipment and technology from those same Northern countries.

Two further factors reinforce this cycle of green dependency. The first is the growing debt crisis across the Global South. Debt crises are nothing new – neoliberalism was built on the enclosures facilitated by the debt crises of the 1980s and 90s. Yet the Global South is facing a debt crisis worse than any of the past hundred years. Huge debts incurred through the COVID-19 pandemic, rising interest rates, the start of the war in Ukraine in 2022, as well as mounting costs from climate disasters have shattered national economies. Governments across the Global South are on average paying out 38 per cent of their revenue to service debts, rising to a staggering 54 per cent throughout the African continent.[23]

The debt crisis is already being weaponised by countries and companies across the Global North. Debt-for-nature swaps are increasing, while Southern governments, desperate to attract investment and capital, are pressured into signing away mining rights at substantial discounts or giving away access to and control over critical infrastructure, such as railways and ports.

Secondly, while major Global South fossil fuel producers have recently reiterated their determination to not only continue to extract but continue to use fossil fuels, demand for oil and coal is beginning to peak. As long as supply exceeds demand, and demand continues to weaken, prices will likely drop over the medium to long term. Major Southern producers often rely heavily on revenue from fossil fuels, creating the conditions for a substantive decline in government revenues, further exacerbating the debt crisis. Worse still, vast amounts of infrastructure, as well as fossil reserves, risk becoming stranded assets, deepening the fiscal crises of Southern nations. All of which means that as climate impacts mount across the South, governments will increasingly be unable to adapt.

The Climate Supercycle, Inflation and the Problem of Undersupply

While steel and automotive workers in Europe fear a global glut of manufactured goods, organisations like the IEA are much more worried about the impacts of scarcity on the transition economy.

There is no absolute lack of metals and resources (aside from water) to underpin the green transition. But there are looming bottlenecks to the available supply of key resources. Over the coming decade most industry forecasts see significant shortages of critical minerals, driving up prices and slowing down the transition.

Crucial to securing future supply will be those 400 new mines. But it's not easy to build a mine. Even with sufficient military force and a supportive government, it takes an average of fifteen years to go from the planning approval stage to production, while it can cost anywhere up to US$1 billion to complete. Popular resistance and conflict with national governments over taxes and costs can further prolong completion time.

Investment is just as crucial to the time and effort required to build a mine. But investment in new mines is scarce. Mining companies are more focused on consolidation than expansion – mergers and buyouts in the sector rose by 75 per cent in 2023 and are expected to continue. Expansion is far too sluggish to meet future demand.

To understand why the sector is focused on consolidation rather than expansion, despite the obvious looming supply crunch, we need to understand the dynamics of resource cycles. Demand for resources is largely cyclical, driven by a complex interplay of economic conditions, technological innovation, regulatory changes and global events. While these all create fluctuations in demand in the short term, over the longer term there are periods of sustained high demand coupled with high commodity prices. These decades-long boom times for commodity prices are called commodity supercycles. The two most recent supercycles have been the post–World War Two commodity boom, and the 2000–14 supercycle driven principally by Chinese demand. The key to these two cycles was an explosive growth in manufacturing output and urbanisation, driving up key mineral prices and demand for fossil fuels.[24] The second supercycle also saw a dramatic increase in the prices of food and agricultural commodities, in part because of rising oil prices.

The end of a supercycle comes as a bust in commodity prices, leaving resource companies with large debts, unsellable inventories and falling prices. The cost and risk involved in expanding extraction means that supercycles are often well underway before companies increase their investments. Investors want their expectations confirmed before committing to new projects. We are presently at the start of a new supercycle, but despite clear signals that demand will increase – guaranteed by government policy – investors are still shying away from new projects, with a few exceptions. With rates of return currently averaging around 7 per cent, many investors see mining as a bad deal.

This won't last, and there are already signs that prices for key transition minerals are beginning to climb. The value of the Bloomberg Commodity Index has doubled since the post-China low in 2015–16, while commodity indexes are at supercycle levels. Oxford Economics, Finch Rating and Goldman Sachs all reckon that we stand at the start of a surge in commodity prices.

There are a few key drivers of this cycle. Urbanisation will continue to increase, while India will likely host the next big construction boom. AI will generate new demands, and China will continue to grow, albeit tepidly. But front and centre is the green transition, making this a climate supercycle.

The climate supercycle will be fundamentally different to previous cycles, however. In contrast to the previous supercycle, this one is not taking place as a part of an intensification of global trade, but in an era of zero-sum neo-mercantile industrial conflicts. The race to secure resources through bilateral agreements and to lock out competitors will generate periods of undersupply of crucial resources. Climate change will make extractivism harder and more expensive, and insurance harder to obtain, as well as periodically disrupting operations and impacting prices and supply. Perversely, as climate impacts mount, there will be increasing pressure to transition, further increasing demand just as supply becomes disrupted.

Additionally, the world is awash in excess savings looking for profitable investments. As commodity prices grow, speculative investments will increase, further exacerbating price volatility and causing havoc – unexpected bankruptcies, delayed projects, or investment in projects that will never work, coupled with rampant profiteering.

The climate supercycle will create a general inflationary pressure, pushing up prices on a wide range of goods and services. It won't be an even pressure, however. Smaller businesses will pay a disproportionately higher price to adapt to the transition economy, as their lower revenues and capital reserves make them vulnerable to higher costs. At the same time, poorer consumers will pay a proportionately higher price for essentials such as food and energy.

Inflation is not a single phenomenon, but a compositive figure used by economists to try to understand price rises across an economy. A typical measure of inflation brings together thousands of price points for hundreds of goods and services to build a picture of price rises in general. In

Britain, the Office for National Statistics puts together a 'shopping basket' of around 700 goods and services to take a snapshot of price rise impacts, called the Consumer Price Index. Central banks and governments will more often use something called 'core inflation', which measures price rises excluding food and energy.

Inflation doesn't affect everyone equally – there is no single 'rate of inflation'. Generally speaking, the poorer you are, the more of your income goes on food, energy and housing. This is as true within countries as between them. In Britain, the poorest 10 per cent of the population had a class inflation rate of 15 per cent in October 2022, while the richest 10 per cent had an inflation rate of 11 per cent. Inflation also differs racially, with non-white consumers facing higher levels of inflation (race is also a bigger factor than class in some countries, like the US). And it is, for the most part, higher in countries across the Global South, especially those that rely on food and energy imports.

Food and energy are excluded from many inflation measures because of their volatility. Due to the impacts of things like speculation, drought, war and natural disasters, their prices tend to move up and down frequently and suddenly. Over the past two decades food prices have become more volatile, as have energy and other commodity prices. Many economists see volatility as something here to stay; with prices for vital commodities such as food and energy prices likely to trend higher for years if not decades to come.

Back during the second peak of food inflation in 2011, the Food and Agriculture Organization of the United Nations set out how this increase in volatility was being triggered by extreme weather events, an increasing global dependence on a few food exporting zones, and increasing energy costs.[25] Volatility has remained high ever since, driving up prices and allowing commodity traders to make huge sums of money speculating on commodity price movements.

It is largely through these shocks that inflation builds. In Florida, a disrupted harvest sends orange prices higher, only for them to stay higher after the damage is repaired, as companies lock in higher returns at consumers' expense. Climate change will drive inflation through an endless series of shocks. And as these shocks cascade throughout the global economy, costs will rise. The long history of European colonialism and industrial capitalism has been one where costs of fundamental materials and key resources fell over time, driving down prices. But no more. The era of falling costs is over, and we are entering one of permanent undersupply.[26]

In many ways, this signals a return to the kinds of political economies that existed before the industrial revolution. Prior to the nineteenth century, one of the main concerns for capitalists and governments was securing a sufficient supply of labour and key resources. These key inputs for economic activity are acquired in two main ways. The first is through a process of extraction, where business and state institutions extract resources from already-existing social systems and environments. For this to work, there needs to be a frontier or 'outside' to continually draw on and expand into, such as uncolonised territories, unexploited environments and the domestic sphere – as well as the ability to undertake the work to profit from expansion.

The second is the transformation and management of key inputs such as land, labour and the environment. After any initial extraction of wealth, capital must take hold of key resources to secure their reproduction as inputs into capitalist accumulation. Where possible, capital prefers labour and nature to reproduce itself at no cost, and for waste to be able to be dumped without incurring any penalties. Yet the dynamics of capitalism tend towards the subsumption of both – the commodification of social reproduction and the waste industry being cases in point.

The productive capacity of capital tends to outrun the reproduction of the resources it can extract from its frontier zones. For workers, this occurs as a series of crises that produce labour shortages, from health and care crises to falling fertility rates, all of which will be aggravated by climate change. With regards to the environment, this manifests as both shortages of crucial resources and the catastrophic impacts of pollution and environmental destruction.

Capital's solution to these crises is to invest in the reproduction of labour and the environment, to try to increase their productivity to match demand and to ensure their reproduction. From education and healthcare to environmental management programmes and tree plantations, both people and nature become subject to increasing management and control. In both cases, the cost of accessing and mobilising these resources rises, generating downward pressure on profits. As a result, while some (but crucially not all) of the shortages produced through the exhaustion of frontiers can be resolved through technological advances, this comes at a price, driving up the cost of doing business. Even if the limit encountered is overcome, it is overcome in such a way as to reduce the potential for future profits and economic growth.

While some limits may be technically overcome, this relies on finding ways to squeeze more out of existing resources, developing new frontiers or finding substitutes. And there are limits to how much more can be squeezed out of existing reserves, as there are limitations to the substitution of one resource for another. New frontiers are becoming hard to find, while those that do exist are difficult to access and expensive to exploit (the seafloor or space, for example). There are also absolute limits on some things such as land and some key ores. Finally, climate change is actively disrupting and destroying both the ecological and human basis of all economic activity, setting the ultimate limit to further resource growth.

At the same time, the despoiliation of everyday life and the destruction of the environment provoke social conflict. People rebel against their impoverishment when businesses and the state try to restrict how much our reproduction costs them. And people struggle against the destruction and enclosure of the global environment, again driving up the cost of doing business.

All these factors create a generalised tendency towards undersupply within capitalism.

Undersupply as a state of permanent crisis emerges out of a specific contradiction within capitalist economies. Capitalism systematically erodes and ultimately destroys the conditions of production: namely, people and the environment. At the same time, productive capacity tends to outrun the supply of key resources over time. The combination of these two factors creates a tendency for the costs of the conditions of production to rise while supply becomes constrained, generating economic crises.

Capitalism relies on keeping the cost of the conditions of production down, in order to ensure a 'reasonable' rate of profit. When journalists or think tanks set out the value of non-waged inputs to the economy, such as domestic or care work, or 'ecosystem services', this cost is essentially what they are trying to measure. It's for this reason Jason Moore and Raj Patel called the conditions of production the seven 'cheaps'.[27]

Climate change takes this contradiction and turns it into a doom loop. Because once climate impacts begin to mount, they begin to undermine the conditions of production autonomously. It is no longer just 'us' directly damaging the world – past certain points, such as 1.5°C, feedback processes take over, meaning changes to the world's climate undermine the conditions for economic production on their own. Even if we could arrest carbon emissions immediately, this process of eroding the basis of our economy wouldn't stop. Worse, as set out across the past three chapters, our current responses to climate impacts will only intensify the problem, leading to worsening climate change, leading to ever more ecological and human devastation.

If stagnation is already producing a widespread sense that the global economy is a zero-sum conflict between trading blocs, one that fuels a revival of protectionism and low-intensity conflicts, climate change will make this the everyday basis not just for international business, but for relationships right across the global economy. Whether between countries, major corporations, small and large businesses, consumers and producers – climate change and the persistence of undersupply will drive a turn to a far more conflictual economy.

Weak economies drive zero-sum mindsets. Growing up in a stagnant economy, especially one shaped by the squeeze, creates a widespread belief that social relations generally are characterised by conflictual dynamics; a belief that you can only gain something if someone else loses it. This is one more way in which the squeeze forecloses possibility, eroding hope for the future.

This brings us back to the problem of inflation. Price rises are not automatic. While newspapers will blithely suggest that the costs of increased commodity prices are somehow 'naturally' passed on to consumers, the reality is that business freely chooses to pass on the cost, to ensure a healthy profit margin. As the economist Isabelle Weber has demonstrated, inflationary shocks are often used to further hike prices and margins, in a process of greedflation.[28] While we can accept that for businesses to remain viable prices would have to increase somewhat (though we might also ask why essential goods should be marketized at all), greedflation suggests inflation is far more of a distributional matter than some kind of natural economic phenomenon.

Inflation is a measure not only of who pays more for what, but, more importantly, who has the power to set prices. Big businesses – especially monopolies – are able to exercise their power to set prices in most markets. This is at the expense not only of consumers but also of smaller businesses, which lack the economic might of transnational corporations. Using inflation as a means of accumulation is but one more element of an economy characterised by monopoly conditions and rentierism. In periods of stagnation, however, we tend to find the mobilisation of inflation as a central accumulation strategy, operating alongside the aggressive use of mergers and acquisitions. Both are predatory practices, well suited to periods marked by low growth and weakened profit rates.

The transition economy is one framed by the need to fix not just the climate crisis but, more importantly for capital, the problem of persistent stagnation. Low productivity growth, faltering returns on innovation and weak investment climates – these are what the transition sets out to resolve. Yet despite significant gains in some areas, the transition is unlikely to lead us out of stagnation. Far more likely is a deepening of the crisis, as the costs of the transition and climate impacts mount. What we will see – what we are already seeing – is a turn to green rentierism and protectionism, in which government handouts and inflation are used to maintain profits within a stuttering global economy.

This Is Not a Collapse. This Is Worse

For years now, Climate Action Tracker has dutifully reported that the gap between government targets of staying below 1.5°C and 2°C and current policy is huge. Little has changed in terms of the projections of where we are headed in climatic terms since the Paris Agreement in 2015. Back then the UN warned that under current policies, we were on track for 3°C. In 2023 the UN repeated the same thing. And in 2024, like clockwork, the UN once again announced to the world that we were on the way to 3°C.

It may well turn out to be the case that the emissions curve bends enough to keep us below 3°C. Early indications are that in China carbon emissions may soon peak, while emissions across the EU dropped by 8 per cent in 2024. Even so, keeping future climate change below 2°C is most likely impossible.

For all the talk of the apocalyptic nature of anything above 2°C, we should be wary of notions of civilisational collapse. Collapse is not coming to save us. Businesses and states are already preparing for a hotter, more chaotic world. Climate activists have long feared a rise in authoritarianism and war in response to climate change; collapse would be almost welcome if it could eliminate both of them for us.

But they can be immensely profitable. The radical fatalism of hoping climate catastrophe will do to the state and to capital what the Left has singularly failed to do is mistaken. Climate catastrophe won't do the political work of revolution for us.

The transition economy must be understood to be a transition into a worse version of the present. It won't end stagnation, nor reverse deindustrialisation or the drain of wealth from the Global South to the North. It won't make for good jobs or better living standards. The squeeze on us will make for excellent returns on investment though.

The contours of the transition economy should make it plain that if we want anything like a future worth living in, if we want more than shattered hopes and precarious jobs, more than endless disruptions and shortages amongst faltering government support, we need to go far beyond calling for more 'reasonable' policies, or for governments to act rationally.

We need to put a stop to their transition.

4

Blockading the Transition

Confronted with the failure of government to do anything remotely adequate to stop catastrophic climate change, and with the abject betrayals of the transition economy, what sets in is a sense of finality, of doom. No longer a fringe politics, the ideology of collapse is becoming a new common sense.

A veritable army of professional optimists persistently decry such doomism, and loudly tell us that if we act now we can still stop climate change before it becomes too dangerous. But unless you are getting paid to be optimistic, you are unlikely to believe that climate change can be arrested before we pass the long-proclaimed boundary between 'safe' and 'dangerous' levels of global warming. Even a moment's reflection here tells us that this has already been passed for millions of people. While some more pragmatic environmental campaigners acknowledge that what we are now fighting for is a 'least bad outcome', this is hardly a goal that inspires dramatic actions and powerful political commitments. Instead, what has emerged as a rallying cry for environmental activists is the idea that we are pushing things so far as to risk the full collapse of global civilisation, and possible human extinction.

While the professional optimists tell us fear is demobilising, the reality is that fear often underpins political action. Organising is largely built on the foundations of our mundane dreads and daily humiliations. It's the transformation of private fears into public concerns that generates the possibility of collective politics. Yet, unlike in workplace struggles, or campaigns to reverse cuts to community services, in the fight against climate change it feels as if there is nothing we can really do except beseech governments to act. It is the fact that climate change is a global problem, implicating everything and everyone, that produces a kind of paralysis.

Climate movements, especially those capable of bringing people into the street to protest, mobilise the fear of catastrophe, tapping into a general sense of dread, but they do so for what seem like quite inconsequential ends. Although leaflets and emails talk up famine and rising seas, the call to act is always a call for street-level protests, invariably demanding that governments 'do something' about climate change.

As we've seen, governments are doing something about the climate crisis. It just happens to not be very effective at preventing warming from reaching 3°C. This is not due to politicians misunderstanding the emergency or being outmanoeuvred by oil companies. The main problem is that most governments are organised around managing the economy, and renouncing fossil fuels would be bad for the economy. And you can't stop climate change without stopping the extraction and burning of fossil fuels.

We don't need to delve into the intricacies of state-theory here to understand the import of this point. Whether or not the modern state is merely a mechanism that manages the affairs of the bourgeoisie, or is a complex set of institutions that comprise a terrain of social and political struggle, it is clearly not the case that governments see their job as using their power to profoundly disrupt economic activity and put trillions of dollars of investments and industries at risk of destruction and closure. And this is setting aside the fact that most major fossil-fuel companies are state-owned and a source of significant national earnings.[1]

Shutting down fossil fuel production and consumption is exactly what it would take to keep future climate change below 3°C of warming.[2] Even managing a transition to an economy that will 'only' produce enough carbon pollution to take us to 2°C of warming is a herculean effort that is already sowing disruption on a massive scale, prompting a spate of social conflicts and political backlashes.

To reduce global emissions enough to stay under 2°C would require not just swapping out blast furnaces and petrol cars for green alternatives, but actively reducing energy and materials use – essentially degrowing part of the fossil economy. The scale of the reductions required far exceeds anything deliberately done at any point in history. The only times such cuts have occurred previously have been in the context of social and economic collapse, such as after the fall of the USSR.[3]

All of this adds up to a political bind. We must halt the extraction and burning of fossil fuels, and we must do so at a very rapid pace. But to suddenly stop threatens to crash the economy and produce severe social unrest, inevitably provoking a broad backlash, signs of which we are seeing already over much weaker forms of government regulation. The only mechanism that appears able to undertake the radical transformation of the economy needed is the state, yet paradoxically the state seems incapable of acting, not least because fossil fuels are currently essential for all economic activity.

It is not fear that produces the broader paralysis of the climate movement, but the seemingly impossible political situation.

Not a System, but a Place

The extraction of fossil fuels is a vast industry. Not only is it a source of obscene profits for capital and a primary source of income for governments, but it also serves as the basis for the global economy – the energetic foundation on which all growth and innovation has been built. And it employs, directly and indirectly, millions of people worldwide. Around 1 per cent of the total global workforce is involved in the fossil fuel industry.[4] Millions more labour in industries that will either need fewer workers after the transition (like steel and electric vehicles) or must be radically downsized (like aviation and fast fashion) in order to reduce carbon emissions.

If we are being honest, any broadly inhabitable future requires ending not just fossil-fuel extraction but a host of industries that cannot exist without fossil fuels and that are profoundly environmentally destructive. Some industries, such as agriculture, will likely require more workers to make them sustainable. Waste management and human services will both require more workers, and whole new industries are emerging that will demand a steady supply of labour. But, in contrast to the existing transition, any serious attempt to stop climate change would invariably mean ceasing some forms of production and shutting down some industries. There is no way to avoid this fact. And while there are many ways to cushion the blow, from reduced working weeks to building up the social economy to strengthening social services and public provisioning, none of these would amount to the creation of an equal number of similar jobs.

Far from a question of a fight between jobs and the environment, however, this is an internal working-class conflict. Stopping climate change is not an agenda being imposed by a professional technocratic elite – it is a question of our survival.

Emblematic of this is a recent showdown at the most senior level of British trade union politics. In September 2024, delegates to the annual Trades Union Congress (TUC) voted narrowly to oppose the British government's plan to ban new licences for oil and gas production. The TUC is the largest federation of trade unions in Britain, representing more than forty-eight unions, which together represent over 5.5 million workers.

The two unions that proposed the motion were Unite and GMB, the second and third largest in the country respectively. They argued that 'while climate change did pose a risk, fossil fuels should not be abandoned until workers knew how their jobs would be protected.'[5] The motion was underpinned in no small part by what one Unite executive member called the 'false promises of green jobs which never seem to materialise'. It was opposed by Britain's largest union, Unison, which argued that while workers in the energy sector should be protected, and investments in green energy made as a part of a just transition, the issue of the ban wasn't only a matter for energy workers, because 'climate change doesn't only impact on energy workers [and] there are no jobs on a dead planet.'

This anecdote is important for several reasons. Firstly, it reminds us that unions are far from a monolith. Many don't see climate change and the green transition as a problem of jobs versus the environment, or even as a threat of job losses at all. There is rarely a consensus inside unions, either. Nor do unions all agree on political positions or economic policies. What is good for some workers might be terrible for others. Sometimes this is obvious – no one on the Left should be under any illusions that police unions are a force for social good. But the same could be said about highly polluting industries, from chemical plants and oil refineries to coal mines. The problem for the Left is that many heavy industries that are environmentally and socially destructive are also bastions of union strength. The Left often confuses the interests of the most well-organised and powerful workers with the interests of the working class as a whole.

This is not a 'forever' argument. Coal miners were instrumental in delivering much of the infrastructure of social democracy. But it cannot be said that, in the face of the existential need to reduce carbon emissions, coal mining is still in the interests of the working class, even if it is in the interest of coal miners. Even putting to one side the other health and environmental impacts of most heavy industries, climate change radically changes what counts as being in the interests of the working class.

Besides, unions are not an unalloyed good. A progressive union must be fought for. Unions are sites of struggle – as any member knows, it is a political project to push unions to take political action. Indeed, just getting your union to take organising seriously and not simply promote pro-business sentiments is a struggle. Unions historically have often backed repressive regimes and laws, supported racist and sexist discrimination alongside wars and imperialism, and opposed socially progressive campaigns. Unions are not inevitably a force for good – like anything else, they must be made progressive by workers in struggle.

The second reason is straightforward. We can't let ourselves be led into thinking that the threat to jobs posed by the absolute need to reduce carbon emissions makes all other impacts less important. Climate change is a class issue, but one for the whole class, not just energy workers. This doesn't mean we should adopt a technocratic perspective and impose change on workplaces; it does mean recognising that this conflict is part of an internal working-class debate that needs to be made explicit.

Justice is difficult and comes with acknowledging that there is more than one kind of transition, and more than one way they can be delayed or slowed down. But there is no delay that could be supported in the name of justice, nor any delay that could be said to be in the interests of the working class. As we speed past 1.5°C of global warming amid dwindling prospects of limiting it to 2°C, and while impacts accelerate and lives are lost, it is clear that the only just transition is a fast one.

The cruelty of a just transition, then, lies in what we would consider to be our greatest asset – that we built and run the world. Which is to say, it is our labour, however coerced and alienated, that is creating climate change. All too often we are told that it is our consumption that is destroying the planet, blaming greed for climate chaos. Or that it is really the fault of a handful of powerful corporations. But the truth is that if we believe in the power of organised labour to change the world, then we must also acknowledge the fact that this power comes from the very work that produces climate change. We made this world – the question is how we can organise ourselves to stop making this one and build another.

Justice should not be confused with jobs. Yet at the same time it is too much to expect workers to exercise their power not just by temporarily withdrawing their labour in strike action, but through a permanent withdrawal from all fossil industries. It is too much to expect workers to lead the abolition of their own workplaces.

Which brings us to the third reason for the importance of the union vote anecdote. While the entire economy certainly needs to change, regardless of what transition takes place, the fight to be had over stopping fossil fuels is not at a systemic level, although gains can be made through legislation and legal action. It is clear that where the fight must be conducted is in those specific sites where the fossil economy is produced, be they pipelines, mines, gas terminals or oil refineries. The fossil economy is produced in specific places. If we are to build the kind of disruptive power that can move us towards the rapid ending of fossil fuels, then it is precisely at these critical junctions that we must organise.

Anatomy of a Blockade

The Left's usual answer doesn't help us. Workers cannot be expected to abolish their own workplaces, certainly not while being squeezed in a time of persistent economic stagnation. To bring an end to fossil fuels we need to turn back to the base constituency of the very idea of climate action: environmentalists.

The environmental movement of the Global North is huge. It is often under-appreciated just how successful it has been, and how large it has become. But the very existence of books like the one you are reading attest to its shortcomings: it has not stopped catastrophic global warming or effectively reduced carbon emissions. Why?

It's often stated that the environmental movement is too middle-class, a point repeatedly underlined by research that suggests that no more than around 20 per cent of most environmental activists are working-class. Putting aside the vagaries of such definitions of class, the argument being made here is that the middle class lacks the power to create deep change because it neither works in critical industries nor is capable of producing a collective, class interest.

Neither are true. The middle class is a huge proportion of the population, and encompasses everything from nurses to engineers to software developers, all crucial not only to the transition economy but to foundational sectors such as healthcare and education. Some of the most powerful industrial actions, and strongest unions, have been within so-called middle-class sectors in recent years. More importantly, the need to stop burning fossil fuels, and therefore to end some industries, starkly reveals the limits of solely focusing on mobilising workers.

Is the issue, then, one of tactics?

Almost all the way across the political spectrum, from policy-orientated green Keynesians to eco-Leninists, the call for a strong state to solve the problem of climate change rings out.[6] Everyone clamours for the state to act, to grasp hold of the problem at a general level. Yet given the manifest failures of policy-orientated activism, and the predominance of symbolic actions within the broader environmental movement, some have advocated more dramatic forms of environmental activism. Andreas Malm's call for the take-up of sabotage, for example, provoked many within the environmental movement, precisely because it was a suggestion of moving past the consensus-driven political strategies of NGO-dominated social movements. Describing fossil capitalism as, in essence, the largest network of infrastructure ever built, Malm advanced the idea that we should shift from protests that enjoin the state to act to direct intervention in this infrastructure, blowing up the pipelines.[7]

Malm is not calling for mass campaigns of property destruction, but rather for existing movements to adopt targeted property damage as one of their tactics. There is an ambivalence in Malm's articulation of sabotage, however, particularly when discussion turns to the environmental movements of the 1990s. After setting out the scale of some of the movements in the '90s, he largely dismisses this history, judging it to be of limited use and unfolding in a void outside of any mass movement.[8]

In fact, almost all these actions took place within the context of mass environmental movements and against the backdrop of internal debates over strategies and tactics. If what we are looking for is the power to disrupt the infrastructure of the fossil economy, to shut down the work of extraction, it is exactly here that we find it. Not in the episodic acts of sabotage, but in how the relevant movement organised itself. There are a huge number of examples to draw on from the 1990s, but two models from Britain stand out: the anti-Genetically Modified Organisms (GMO) campaign and the anti-roads protests.

GMOs are plants modified by having new DNA inserted into their cells. GMO development uses biotechnology to change the fundamental characteristics of a plant, often by introducing DNA from an unrelated organism. Concerns about the introduction of GMOs into Europe first emerged in response to GMO foods turning up on supermarket shelves in the mid-90s, and later in reaction to the trialling of GMO crops on experimental farm plots. Large environmental groups such as Greenpeace and Friends of the Earth raised the alarm about the potential risks of GMOs and campaigned against their introduction, sparking public controversy. But it was the direct-action campaigns of Earth First! and Genetix Snowball that put paid to the effort to introduce GMO crops into British agriculture.

The two groups organised direct actions against the GMO field trials – both overt and covert attacks, where activists would descend en masse and pull up the GMO crops, bagging them as biohazards. These acts of sabotage often took place in full view of the police and media, against a background of growing public support. Genetix Snowball took to organising direct actions every second weekend, managing to attack a significant number of the 300 sites around Britain where genetically modified crops were being grown at the time. It was the systematic nature of the disruption, coupled with mounting public anger, that ultimately forced GMO companies to pull out of the trials. Even as arrests and criminalisation escalated, actions continued. Major supermarkets and food corporations also announced that they would not sell or use GMO crops as ingredients, adding to the pressure.

GMO companies, facing overwhelming resistance and commercial uncertainty, largely abandoned their plans for GM crop introduction, contributing to the eventual de facto government moratorium on the commercial planting of GM crops in Britain. Similar protests were taking place across Europe, with France in particular hosting a similar direct-action campaign, leading to the eventual ban on GMO crops across several European countries and stringent controls at an EU level. After Brexit, new regulations were adopted allowing for some limited forms of GMO crops, yet by and large the de facto ban on GMOs remains.

Before the anti-GMO protests of the late 1990s, Earth First! had been involved in another mass campaign that also used direct action to successfully derail destructive development schemes.

In 1989, Prime Minister Margaret Thatcher launched a ten-year road development programme that aimed to construct over 4,000 km of roadworks, including 150 new bypasses. Campaigns against the roads, many of which would destroy woodlands and homes, started in 1991, with some of the first direct actions taking place in 1992 at the Twyford Down construction site. Machinery was occupied, roads blocked and protest camps set up. While these actions were ultimately unsuccessful, it was just the first of many such site-specific protests. By 1995, 300 projects had been cancelled. That year a huge protest was mobilised against the building of the Newbury Bypass, with thousands taking part. Tree-sits (platforms in trees hosting activists) and tunnels were built along the planned route, blocking construction. As with Twyford Down, the Newbury Bypass was eventually built, and at great cost, but just two years later – after endless delays, cost overruns and project cancellations – the newly elected Labour government cancelled the rest of the programme. Once again, direct action had got the job done.

Within both protest cycles, as with many others throughout the 90s, acts of sabotage took place. We could argue that the entire anti-GMO movement was one of mass sabotage. And while both movements hold many lessons for us, not least the need to contest facile narratives of progress and development and to de-fetishise new technologies, the main one is how we can go beyond demanding government action and stop things ourselves.

What gave both British movements their power comes down to their ability to cause sustained disruption in the operation of something, be it a GMO trial or a road-building programme. As with strike action, power came from the damage that could be inflicted. And this damage was not just a consequence of the tactic of property damage or direct action, but of the ability to sustain the campaign. Power comes from being able to make the protest endure.

Sabotage is just one aspect of direct action – political action that seeks to actively and directly disrupt something: the workings of a business or factory, a mine or farm. It aims to stop something happening and is deliberately antagonistic. The environmental movement has a long history of taking direct action – against the construction of dams or roads, against logging or mining. And this history is one of innumerable victories.

In one of the largest studies of contemporary direct actions, it was found that over a quarter of what the authors called 'place-based movements' were successful, with projects being 'shelved, suspended or delayed'.[9] From victories against fracking in England, to successful pipeline struggles in the US, to campaigns against mining in Serbia, to resistance to fossil-fuel extraction in Ecuador and Bolivia, direct action has a strong track record in achieving actual change.

It's easy to list the historical examples of this. But contemporary actions are harder to find. Examples such as the #NoDAPL pipeline blockade in North Dakota, or the anti-reservoir movement in France, are unfortunately rare. More often what we see within the environmental movement are actions that produce short-term disruptions, such as occupying a building or blocking a road, often for only a few hours. Even the most spectacular actions from the past few years, such as those against coal mining in Germany by Ende Gelände, are little more than short-term performances that largely aim to shape public opinion. In reality they are what might be called militant stunts, of the sort that UK-based Extinction Rebellion specialise in. These have made their mark on public opinion, and not always in a positive way, but they do not constitute direct action in the most powerful sense of the term.

Direct action is not a stunt but a sustained disruption. It works because the disruption causes the mine, construction project, road-building or logging to stop. What makes the difference in the end is that direct action is sustained. It lasts as long as it needs to: often months, sometimes years. Many of the most significant victories of the global environmental movement in the past have come from sustained direct action.

But for that threat to be real, for direct action to be a threat and not a stunt, it must be able to be conducted at length. A direct-action campaign depends on organisational capacity to sustain itself.

If the environmental movement is to become a genuine threat to fossil capitalism, and if we are to have any hope of staying below profoundly dangerous levels of climate change, we need to build the movement's disruptive capacity. The way to do this is not through any one demographic or tactic, but by constructing a machine capable of creating the ecological militancy and cadre we need to sustain disruption. That machine is the blockade.

A blockade is not just a means of physically stopping the flow of people, machinery or resources from entering or leaving a site. It's more than barricades, people locking onto machinery, roadblocks or tree sits. It's a means of sustaining opposition. It's an organisation that not only shuts something down but produces the ability to sustain that shutdown.

Let's take a successful example of a blockade – the campaign to stop the creation of the Jabiluka uranium mine on indigenous land in northern Australia. Twenty years ago, Energy Resources of Australia (ERA) set out to build the first of twenty-five proposed uranium mines. The opposition campaign began with the resistance of the Mirarr, the traditional Aboriginal owners of the land, and soon gathered the support of a wide range of regional and national environmental groups. Through these groups, but primarily through student environmental activist networks recruited by the Mirarr people, a system was created to recruit, train and transport more than 5,000 people to the protest camp set up near the proposed mine site. The camp lasted eight months, with daily actions and continuous blockades, leading to over 530 arrests. At the same time numerous protests, rallies and actions took place around the country, including another two-week blockade of the ERA offices in Melbourne.

As the wet season approached, the camp was lifted, having substantially transformed public opinion against the proposed mine. ERA was in dire financial straits, and there was now a credible threat of the blockade being reinstated after the wet season. The Mirarr people continued to apply political and legal pressure, and the mine was delayed, along with the other planned projects. In 2002 the mining company Rio Tinto, the new owners of ERA, announced that Jabiluka would not be mined without Mirarr consent.

The victory of the Mirarr people and the Jabiluka cam-paign tells us so much – how to work as allies, how to blend campaign tactics, how to build public support for illegal actions. But the main message is how to build environmental militancy, skills and competency amongst activists, and capac-ity within a movement. Blockading isn't just a powerful tactic or strategy – it's a method for building movements and power.

Different actions require different forms of organisation. The current emphasis in Britain is on militant stunts and rallies. Both focus on generating pressure on government through publicity. Recruiting people and training them to undertake militant stunts, and creating the infrastructure to undertake them, builds an organisation that prepares people for short-term action. The focus on rallies supports building professionalised structures that work through networks of existing organisations and are centred on 'getting the word out' and training for crowd control and press liaison. Nei-ther approach looks to train and support militants who can take action indefinitely.

In contrast, we can think of the blockade as comprising four key elements: recruitment, training, planning and infra-structure, and permanence.

Recruitment comes first, and always remains a necessity of the blockade. This is because the blockade must be per-manent. Even if it only lasts a week, it has to be understood as permanent because it needs to be built to last until it wins. Blockading calls for an organisation that has a dedi-cated recruitment function, one that could ensure continuous expansion and engagement in order to commit as many people as possible to training and struggle.

Training is crucial: not only in the technical skills of blockading but in how to confront the police, how to deal with security, how to maintain a camp or set up a blockade. The psychological and social aspects of training are more important here than the practical skills. Preparing people to persevere over the long term, and work as a collective, is the cornerstone of training. To that end the blockade comprises a whole host of schools, training sessions, materials and processes to create trainers themselves.

All blockades are physical spaces, making planning and infrastructure vital to success. We need to plan the creation and maintenance of blockades as forms of infrastructure; we need the materials not only for a weekend camp but a season-long encampment. Planning for the long term is different to planning a one-off event or action. The crucial difference is that for the one-off event we centre the safety of participants; for the blockade, we put the question of how to reproduce people and the blockade itself at the heart of our planning. Planning and infrastructure carry people through the campaign, building the strong physical basis blockaders need to endure.

Lastly, the blockade is permanent. The political horizon of the blockade isn't to secure a shift in public opinion or government policy, but to make the blockade last long enough to physically stop something happening. This means people are recruited and trained, and organisations plan and organise for the long haul from the beginning, on the assumption it will never end. But it means more than this – the blockade as a way of organising never ends. No one else is coming to make the change. We organise and recruit and train assuming each blockade is just one small part of the larger disruption we need to create.

All of this creates a specific kind of community of resistance, one that supports and cultivates a continual process of direct action and militancy, and embeds direct action in a broader tapestry of mobilisations, actions and support. It also enables a direct engagement with affected communities, workers and, in settler colonies, indigenous activists and communities – either by supporting their struggles for justice and land rights, or by the necessity of confronting the violence of colonialism that arises from any political activism on stolen land.

The scale of the climate problem terrifies us. It pushes us to look for quick answers or magical actions that can solve everything now. Organising is hard, and often slow. Conflict can't be avoided, and there is a real discomfort and unsafe edge to putting yourself in harm's way.

But we need to embrace this discomfort if we are to win. The blockade is our most effective approach. Stunts, marches and petitions can all effect change, but only if backed up by the very real threat of sustained disruption. We need to focus on taking what we already have further – our networks, campaigns, actions and collectives. We need to look to specific sites and battles not as the solution to our impasse but as the key to building the movement that is indispensable. The blockade is our political horizon. Towards it we must turn, and through it we must go.

Beyond Pipelines

But what are we blockading?

It could be said that the obvious answer – fossil-fuel infrastructure – is untenable. But we have to stop burning fossil fuels to arrest climate change and keep future warming below 3°C so we need to go after this infrastructure. It's considered untenable not because of the ambition – ambition should not scare us. It's considered untenable because it would produce massive socio-economic disruption if successful.

Alarming as that sounds, if we are being honest that is exactly what is called for. To advocate for less is really to advocate for a 3°C future.

One might think that 'everyone already knows' what this would be like, and yet it's worth spelling out, because more often than not the impacts of future climate change are discounted or minimised in the present, with the temptation in the Global North being to assume the consequences will affect someone else and not 'us'.

A 3°C warmer world would be marked by severe and escalating climate disorders that would significantly reshape life on Earth. At this level of warming, heatwaves would become extreme, with parts of the tropics becoming uninhabitable for humans. Crop yields would decline sharply due to droughts and heat stress, threatening global food security. On a regular basis, food production would collapse in multiple breadbaskets. Sea levels would rise by over a metre, perhaps as high as two this century, displacing millions from low-lying coastal areas, while the frequency and intensity of hurricanes, wildfires and floods would increase dramatically. Many ecosystems, including coral reefs and the Arctic, would face collapse, and the likelihood of crossing tipping points – such as the irreversible melting of the Greenland ice sheet – would become dangerously high, pushing the planet towards even more catastrophic changes.[10]

Faced with these prospects, any environmental movement must choose militancy over political realism. Such militancy must be grounded in broader social and political movements, for otherwise disruption risks producing a void to be filled by reactionary political actors. But it will, by design, be disruptive. We can no longer pretend that taking action to stop climate change could be anything else.

To start with, strategically we need to distinguish between the fossil-fuel infrastructure that is in the planning and construction stage, and that which already exists. From this perspective, it is simply easier to blockade the proposed site of new fossil infrastructure, or the construction of it, than already existing and operating infrastructure. Blowing up functioning oil and gas pipelines would cause monumental ecological disasters. At a political level, blockading new fossil infrastructure directly contests the transition that relies on it.

Yet the focus of our strategy shouldn't be limited to upstream aspects of production – wells and mines – although, as is the case with mines, the fact that they are profoundly destructive of local environments and communities often inspires strong local opposition campaigns that are a natural constituency for environmental movements. Rather, as with many other aspects of contemporary capitalism, fossil infrastructure is organised through choke points.

A choke point is a strategic location where the flow of goods, resources, information or people is significantly concentrated and can be easily obstructed or controlled. Choke points are often targeted in conflicts, precisely because they represent vulnerable, high-leverage sites in complex networks of trade and production. Fossil-fuel choke points are strategic locations in global and regional energy infrastructure where the extraction, transportation and distribution of oil, gas, or coal is heavily concentrated. These points can include oil shipping routes, pipelines, refineries and natural gas terminals, where a significant portion of fossil-fuel supply passes through narrow, highly trafficked corridors or facilities.

While it has been argued that we live in an age defined by declining levels of workplace struggle and weakened unions, and thus in a historical period marked by the centrality of the riot or social strike as a weapon of class war, choke points have historically been favoured terrains for workplace organising.[11] Dock workers and coal miners particularly, operating at critical junctures, have been able to use their leverage to exact huge gains, both within the workplace and across the social field.[12] But targeting choke points is not solely the preserve of workers within their own workplaces. In 2000, angry at rising fuel prices, British farmers and lorry drivers blockaded oil refineries. After farmers blockaded Stanlow Refinery in Cheshire, the blockades spread rapidly, and within three days six of the nine refineries and four oil distribution depots in Britain were blocked. Panic buying and rationing ensued, and twenty-four hours later around 3,000 service stations closed for lack of fuel. Food in supermarkets began to be rationed, and hospitals cancelled non-essential operations. Five days later, after the intimidation of drivers and the deployment of military personnel, the blockades began to wind down and supplies were restored. By this time similar protests and blockades had spread across Europe.

What the UK refinery blockades show is just how vulnerable fossil choke points are. There are only six refineries operating in Britain, while France has seven. In the US there are far more – around 132. The scale of US fossil infrastructure is vast, as befits the world's largest oil and gas producer, and a country the size of Europe. While Europe has a total of fifty-seven LPG (petroleum gas) import terminals, there are more than 170 LNG (natural gas) facilities operating across the States, including seven export facilities. But this does not make US infrastructure any less vulnerable, as most of the LNG production is for export, turning the export facilities into vital choke points in the global fossil fuel economy.[13] Similarly, while there are a large number of oil refineries, there are only a third more than across Europe, and a significant portion of refining capacity is concentrated among a relatively small number of companies and facilities.

Strategic decisions over which facilities to disrupt need to be coordinated as much as possible with affected communities and unions. And it makes far more tactical sense to focus on a smaller number of high-impact sites – European refineries, US export terminals, or the large coal basins that supply the bulk of Europe's coal. Equally crucial are proposed facilities, such as the 133 new natural gas-fired and four oil-fired power plants that investors are supporting in the States, in no small part to power the exponential growth of data centres;[14] or the new mine sites and productions fields mooted in Australia or Canada.

Some will warn that to target such critical infrastructure would be to invite harsh repression. Yet environmental defenders are already being murdered, largely with impunity, around the world, and far less disruptive protests face brutal police repression. It must be recognised that any protest or action worth the effort of organising will be met with political repression. The whole of the Global North has been steadily militarised, with governments broadly taking an authoritarian turn as part of the punitive austerity imposed after the 2008 financial crisis. From Black Lives Matter to Extinction Rebellion to marches against the Israeli genocide conducted with Western support in Palestine, political repression is an unavoidable aspect of political action. But this is not a reason to retreat from action; being ineffectual is no defence from repression.

Major sites of emissions are not confined to oil and gas infrastructure. Airports (and their expansion), industrial livestock megafarms, data centres that power water- and energy-hungry AI systems, power stations and industrial hubs: all are vital sites when considered as sources of climate-wrecking pollution. Different countries and regions will have different priorities. For example, in Britain some of the key sites that produce climate change include chemical manufacture facilities in Runcorn, Cheshire, as well as industrial clusters such as those in Teesside in the Northeast. These and other sites could and should be part of a community-orchestrated transition, even if such a transition involves an overall reduction in production, as it is a necessary element of any real action on carbon emissions. But no such case could be made for livestock megafarms or airports, neither of which can operate sustainably in any meaningful way.[15]

A core part of any alternative just transition must be debate over what technologies to pursue. It is far from given what technologies we actually need for a just transition, and which technologies merely facilitate new rounds of green capitalist accumulation. As such, the development of green industries may also become a necessary target.

New infrastructure and technologies form a vital part of the existing transition plan. From wind farms and new power cables through to hydrogen production facilities and EV charging stations, the transition is a world-historical buildout of new systems and materials. Some of these new pieces of infrastructure are crucial to any plans for a transition, be they business plans dreamed up by CEOs in boardrooms or our plans generated from below. But that doesn't apply to all of these projects. And even in the cases where they are necessary, it is likely that, for us to have a good life, they aren't needed at anywhere near the extent or scale required by capital.

At the same time, questions remain as to who and what should be sacrificed to make the green transition. The buildout of green capitalism will involve a huge surge in mining and extraction, all of which takes place somewhere, usually at the social margins. We can expand the meaning of extraction here to include renewable energy production, which is often situated in marginal communities. From lithium mines in Argentina to wind farms on Sami pastures in northern Norway, we need to defend the right of impacted communities to refuse development in the interests of someone else's green economy.

We must denaturalise the transition and ask if what we want is the 'greening' of a destructive form of consumerism that requires the sacrifice of countless peoples and environments, mostly in the Global South, or whether our transition could not in fact do with less in order that others may live more. Reducing the impact of the Global North's way of life – its imperial mode of living – is a necessary part of any just transition.

To that end, not only should many of the projects the powerful claim are a necessary part of the transition be opposed, but we must incorporate long-neglected practices of international solidarity into our blockades. We must join campaigns against pipelines that fuel Israel's war on Palestine, against the construction of mass solar arrays in North Africa to power European homes or against the proliferation of mines throughout the Congo to build our batteries. The blockade cannot stop at the borders of the Global North.

Many of these projects of green development are already sites of fierce resistance. What I advocate here isn't novel or new. What I am arguing for is a scaling-up of our ambition.

In other words, the blockade is not just a means of stopping the fossil-fuel economy, or building a movement of militants, but forms part of a broader process of political education. If our struggles are to go beyond the purely defensive, then they must be grounded in explicit political frameworks and critiques. Ultimately the blockade as a form must be subordinate to a greater political project and vision housed within the networks and organisations of a community-led transition.

This will not exempt us from conflict within our networks and movements. There is no conflict-free transition, and faced with accelerating climate change there are no easy solutions. Things must stop; industries must shut down; whole forms of fossil-fuelled life must be abandoned. Justice here does not mean avoiding the hard choices, though. What it does mean is politicising them, making them a problem that we can collectively decide on. For too long the environmental movement and the Left have been afraid to honestly address what is necessary in the transition, and all it would imply. Revolution is an often-used word that hides the historical reality of profound disruption. But the blockade puts that most direct question of the transition to us: what must we stop in order to live?

5

Refusing the Price of Their Transition

At the peak of recent food and energy inflation in 2022, protests and riots shook Europe. Italians burned energy bills in coordinated actions while the French took to the streets to protest the crushing cost-of-living increases. There were demonstrations across Germany and Spain, while in Britain protests were buttressed by the emergence of the Don't Pay campaign – an organised mass refusal to pay energy bills that grew so large, so quickly, that the energy company E.ON declared it an 'existential threat'.

While the protest wave of 2022 was quickly repackaged as part of a 'climate backlash' narrative by right-wing politicians and conservative journalists, in reality it was part of something else: a global movement pushing for price controls from below. Sure enough, from Hungary to Brazil, from India and Argentina to Turkey and Egypt, numerous countries rapidly introduced price controls on food and energy bills to ward off social unrest and placate protesting crowds. Initiatives such as subsidised or free public transport, tax relief and social subsidies were put in place. Price controls became commonsense policies.

The year 2022 marked a milestone in people's refusal to let their community or workplace be sacrificed on the altar of crisis or in the name of the green transition. They refused to accept the loss of land or work, or to just 'be poorer'. These struggles are all part of the war of transition – moments of social conflict over the emerging social 'deal' that will form the basis of our expectations, aspirations and living standards within the transition economy.

This violent reordering is made to appear natural or inevitable. The climate crisis will make us poorer, and to prevent it being even worse, 'we' need to make sacrifices.

No one doubts that life will have to change. We have little choice now but to begin to adapt to worsening climate impacts, and focus on transforming how we work and live. But there is nothing natural or inevitable about how this version of transition will unfold. There will be a squeeze; but that does not mean we should be the ones to suffer.

The powerful demand that we accept being poorer and more insecure in order to limit the exposure of capital and the state to climate losses. There is nothing secret or obscure about this. Everyone who pays attention already knows it. As with Port Talbot, the choice is presented to us as an ultimatum. Either we accept it or capital flees and the state retreats, leaving us exposed to worsening climate impacts and economic ruin.

This threat is present every time we buy food or think about putting the heating or AC on. The squeeze is administered through prices, making them the frontline of the climate struggle in our everyday lives. And while fear of popular discontent, or crude electoral calculus, can lead governments to institute (often merely temporary) price controls, we can't rely on the state to limit the effects of the climate squeeze. We need to refuse to pay their price and set our own.

The People Are Disappointing

Popular movements forcing limits to food price inflation through mass protests and riots are likely not the first kind of refusal that comes to mind within the war of transition. One thinks rather of farmers blockading roads to protest new pollution restrictions, or petty landlords in Germany fighting against heat pumps that would erode their rental empire incomes, or even far-right conspiracy theorists rioting against fifteen-minute cities, imagined by them to be a secret world-government plan to institute 'climate lockdowns'.

There is as yet no clear political organisation of the discontent about the squeeze on our lives. It is uneven, prone to polarisation and inchoate. Examples of radical or progressive organising efforts to give shape to the fightback against the squeeze are all too rare. Worse, many on the Left and in environmental movements fail to see the organising potential within the rising dissent.

The problem is that because the squeeze is administered through prices, it appears to boil down to a question of choice. Either you 'choose' to pay for the heat pump or electric car, choose to retrain as an installer or waste management professional, or you are not taking climate change seriously. It's your income vs the environment: your choice.

While it's clear government and business aren't doing enough to arrest climate change, it increasingly seems as though people aren't willing to make changes either.

The financial press is already wondering whether electric vehicle sales have begun to peak. Forbes described the slowdown as the market 'hitting a wall of consumer ambivalence'.[1] Researchers have begun to look into the psychology of adoption, hoping to find some secret sauce that would give sales a boost. After years of growing popularity, meat alternatives, too, have begun to lose customers. Notwithstanding the torrent of articles on veganism and plant-based foods, people have begun to turn their backs on going meat-free, with many retail commentators claiming the market is over-hyped.

By now, everyone has seen a version of the 'what can I do to reduce my carbon footprint' list. From meat-free Mondays to installing solar panels and buying green, to offsetting flights and foregoing fast fashion – the list of individual 'choices' is both long and clichéd. But for the most part people not only refuse to consume less, they also tend to not buy green products, even when they say they want them. The paradox of the elusive green consumer haunts most big brands and companies, who claim they would love to go green but the demand 'just isn't there'. This is not strictly true. Still, there seems to be a ceiling on sustainable consumer purchases, where only so many people (around 30 per cent in most cases) will pay a premium for greener products, or even make the switch at all.

Consumer behaviour is a knotty problem for the transition, and a repeated concern for politicians and policymakers. It is a problem in so far as a huge amount of future carbon-emission reductions will supposedly come from us changing what we buy and how we consume. The IEA in its 'Net Zero by 2050' report, sets out how consumer choices and behavioural changes are to be responsible for driving close to 40 per cent of all future carbon emissions reductions.[2] So far, people aren't sufficiently choosing sustainable options or, in the case of high-carbon products such as international flights, choosing to consume less at all. There is a 'green' demand problem.

Some environmentalists decry the failure of people to choose better, and declare them to be a plague, the destroyers of the world, arguing that the fundamental problem 'is us'.[3] The idea that what people want is to consume acts as a brake on government policy; what government would voluntarily take burgers and cheap flights away from voters? The idea that all the working class desires is SUVs, and bigger houses with more stuff, fuels a strain of anti-green leftism that pins its hopes on solving climate change – and simultaneously placating the working class – with fantasies of space mining and boundless socialist technological development.

The assumption is that the human desire to consume is boundless. Ultimately, the choice we seem to face is for government to legislate the right consumer choices through taxes or bans, or to indulge in fantasies of technological salvation.

The problem with this assumption is that it's wrong.

Consumption, as a monolithic category, is strange.[4] Buying necessities like food or paying utility bills is made to look the same as buying luxury suits or handbags. Talking about all purchases as consumption makes them all seem like the same kind of choice, much in the same way as talking about the jobs market turns the kind of job you do into an expression of your desires.

The average person in the Global North spends around two-thirds of their income on necessities (the figure is much higher across the Global South). If we add in 'discretionary' items that are really no such thing – phone contracts, work clothing, internet bills – we spend up to 70 per cent of our regular incomes on the things we can't do without.

Almost all this spending has profound ecological impacts; but consumers don't 'choose' these impacts. These are not things people can do without. Not only do they have no say in how they are produced or structured, often their choices are scripted by budgetary constraints or, just as frequently, by marketing and the mobilisation of the architecture of desire by supermarkets and retail businesses. The freedom to choose is often a misnomer.

The neoliberal myth of choice underpins the growing political consensus that there are hard limits to how much change people will accept as a part of the transition. Attempts to enforce the adoption of heat pumps has, to date, just facilitated the electoral rise of right-wing political parties. Programmes to reduce car use, or to move people to EVs, provoke protests and electoral losses. Talk of carbon taxes or green levies just turns people against the green transition and towards new forms of climate denial.

At this point, it is important to note that this growing consensus is emerging in response to specific events, from farmers' protests in Europe to anti-'climate lockdown' campaigns by conspiracy-supporting far-right organisations. Of all the recent backlash protests, however, it is one specific insurgency in France that forms the foundation of the myth that people won't accept the need to change their behaviour as a part of the transition.

The Gilets Jaunes (Yellow Vest) movement erupted on 17 November 2018 as a protest against a proposed fuel tax increase, quickly escalating into a nationwide uprising. Hundreds of thousands blocked roads and roundabouts, while demonstrations took place across French cities, leading to huge riots, including in Paris. The protests soon forced the government to suspend the tax increase (in early December), but by then the movement had already mutated into a broader rejection of President Macron's economic policies.

Contrary to right-wing political fable, the Gilets Jaunes were not simply 'anti-climate'. The movement continued to fight well after the tax rise was rescinded, opposing Macron's entire neoliberal market reform agenda. What marked the Gilets Jaunes movement was the heterogeneity of its composition. Struggling middle-class workers, public sector employees, service workers and shopkeepers, farmers and industrial workers. The revolt drew broadly from across French society, where a common sentiment of *ras-le-bol* (being fed up) had taken hold.

The Gilets Jaunes opposed policies that placed the burden of the energy transition on ordinary consumers rather than on industry. Ultimately, the movement expressed a demand for a fairer distribution of wealth, time, and hope, rejecting the costs imposed on the working class by government and corporate interests.

The refusals of the Gilets Jaunes, like the protesters burning their bills in Italy, or the demonstrators calling for price caps on basic foods, are not symptoms of a vast anti-green groundswell or a working-class rejection of green measures. They are a refusal to be drawn into a new social bargain where our lives are reduced to nothing more than survival.

The characterisation of lifestyles in the Global North as profoundly ecologically destructive often leads to either moralistic condemnations of consumers or, conversely, a rejection of any sense of complicity. It is incontestable that life in the Global North is founded on an imperial mode of living.[5] The consumption patterns and lifestyles across the countries of the Global North rely on the exploitation of resources, labour and the environment from the countries of the Global South. Yet the idea of choice, so crucial to the mythology of the market, turns this economic structure into the product of individual choices. Which is precisely the wrong way around.

The organisation of the global economy as a vast imperial machine has not been undertaken to service our needs. Its expansion into every aspect of our world and lives has been deliberately designed by capital in order to make it impossible for us to meet those needs outside of the market. The conquests and violence of the imperial mode of living include the spaces of our everyday lives, down to the colonisation of our desires and attention. Even our hopes have been captured by carbon-intensive industries that perpetuate the destruction of countless lives and ecosystems around the world.

The illusion of choice masks this compulsion. Yet, with few exceptions, since all our choices are destructive, there seem to be few virtuous options. At best we can 'choose' to consume something slightly less destructive, slightly less oppressive. But we are still compelled to choose something. There are no choices outside of this system. Given the hold the myth of choice has on our imaginations, people in this articulation can only ever be disappointing.

This isn't to say these things don't need to change – they do. Nor that we are not unwittingly complicit in the Global North's imperial violence – we are. It is, however, to insist that it is wrong to see consumer choice as the key mechanism of transformation. The shift to taxation and bans is in many ways a recognition of the limits of the mythology of choice. But it is a recognition built on the wrong foundations. It is not people's excessive desires that are the problem, but how their fundamental needs are organised through a system that relies on constant consumer spending and compound economic growth to survive.

There is no shortage of ideas for how to solve this problem by transforming the basis of our mode of living. From neighbourhood canteens serving subsidised or free meals, to the massive expansion of public transport and bike infrastructure, to universal basic services, rent controls and shorter working weeks coupled to longer holidays – most people's lives can be made better and more sustainable without increasing taxation or banning everything, but instead through the creation of forms of public luxury and provisioning.

As it stands, the current transition to a low-carbon economy is not a utopian plan. It's not even a social democratic plan. It's a business plan. It's a plan to ensure that capital isn't required to pay the cost of the transition. So why would they start giving away things for free now?

Against the Price

The squeeze is a distributional struggle, where business is using the shocks and disruptions produced by the climate crisis to increase profits and pass the costs of the transition onto us. Prices are not a natural market phenomenon, but the product of class struggle.

The drive to accumulate through inflation is not uncontested, however. While we are told we will have to accept paying more for things and will be poorer overall, growing numbers of people are refusing to accept this new 'standard of living'. Over the course of a few months in early 2023, the Scottish activist group This Is Rigged set about stealing food from major Glasgow supermarkets and giving it to the public. Starting with bread and roses nicked from the super-market chain Tesco and then handed out on the street in front of the store, the group declared:

'We are asking for the bare minimum: our inalienable human right to baseline food security; but we are also asking for dignity. No longer will we beg for scraps from the tables of greedy CEOs. No longer will we be at the mercy of cor-porations who price us out of eating on a whim, dependent on them to provide us with our foundational needs.'[6]

Less than a week later the group targeted fifteen super-markets, taking potato scones to distribute around the city. Shortly after that they moved onto baby formula. The actions, part stunt, part actual theft, were coupled to two demands: that supermarkets slash their prices to match March 2021 levels, and that the Scottish government fund one community food hub for every 500 households in Scotland.

Britain is in the middle of a shoplifting epidemic. According to the Office for National Statistics, 2023 reg-istered the highest number of thefts on record – more than 430,000 cases.[7] This is likely an underestimate, with the British Retail Consortium reporting over 16.7 million thefts between 2022 and 2023.[8] The Consortium's figures would suggest shoplifting is an everyday act for many, with around 3 million people shoplifting in any given year. In the US retail theft is also surging, by 20 per cent per year in some states, with total annual losses amounting to over US\$112 billion.[9] While precise figures are hard to find, at least 10 per cent of Americans shoplift at least once in their lifetime, with the true figure likely to be substantially higher: at least one recent survey showed that one in five Americans have done it.[10]

Shoplifting and theft have cohered into mass movements in the past. Perhaps the sharpest account comes from the historian E. P. Thompson, who described how communities in pre-industrial England would often fight back with protests, riots and other forms of direct action when prices rose unjustly or essential goods became inaccessible.[11] Thompson argued that these actions were not random or chaotic, but rooted in a shared moral framework where communities held a collective expectation that essential goods, particularly food, would be fairly priced. When these expectations were betrayed, especially by profiteering or manipulation by merchants, people would organise demonstrations or seize control of goods to enforce fair pricing. Such riots were a profound collective refusal to accept the degradation of standards of living through pricing mechanisms. The defence of 'customary entitlements' was fundamentally a defence of a grounded understanding of justice, held in common throughout the community.

This same sense of enacting justice by contesting prices powerfully re-emerged during the explosive inflation of the 1970s. The revolt against the austerity and violent control exerted by prices reached its apogee in the Italian Autonomia movement. Autonomia represented a radical departure from traditional leftist politics, and was driven by a diverse coalition of workers, students, feminists and unemployed youth bound together in their refusal of existing workplace structures and mainstream politics, including the Communist Party. It was a rejection of institutionalised politics, in favour of direct action and workplace-based organising. The refusal of all mechanisms of control became a pervasive attitude and practice, and the period was marked by intense political conflict and frequent clashes with the state.

During this wave of protest and dissent, theft was transformed into a mass, organised practice of *auto-reduction*. Political collectives would organise mass 'proletarian shopping' actions, assembling people at a specific time and location before descending on a supermarket. Once inside, activists would take control of the PA system and announce, 'Today everything is free! Don't pay! Don't pay!' often sparking an organised looting of the shop.[12]

Inflationary shocks engender a sense of injustice as well as misfortune, making clear our absolute dependence on the market. Inflation is not the only way in which we experience this dependency, though. Austerity policies enacted after the 2008 crisis also prompted a wave of riots and protests, most dramatically expressed in the civil conflicts and incomplete revolutions that shook North Africa and the Middle East.

As the squeeze develops and intensifies, contestation of the contraction of our lives will spread. Riots, protests, occupations, thefts and refusals will become daily events, barely remarked upon. Against the foreclosure of possibility, the squeeze will provoke severe unrest.

Don't Pay

By mid-2022, 7.8 million people in Britain were in debt to energy companies – almost one in four households. The total energy debt in Britain reached approximately £2.5 billion. Meanwhile millions of other households were mired in fuel poverty, and spending more than 10 per cent of their incomes on energy bills.

By October of that year the government had implemented a price cap on energy to households, shortly after energy giant E.ON declared that the risk of non-payment constituted an existential threat to the company and the broader sector. While the price cap on energy didn't go far enough, and left large numbers exposed to significant price increases, it was nonetheless a significant moment – especially as the government that implemented it was the most neoliberal regime of the past twenty years. What forced the viciously pro-market government's hand was the Don't Pay campaign.

Launched in June 2022, Don't Pay started as little more than a website and a small organising group. The campaign called on people to pledge to refuse to pay their energy bills from 1 October, unless government and energy companies took action to reduce the cost of energy. The pledges to withhold would not be 'activated' until there was a critical mass of pledges, the initial target being set at 1 million. Through a combination of strong media work, the building of local groups able to promote the campaign through fly-posting and local meetings, and a specific programme of training local organisers, the movement grew rapidly. By August, it had 1,300 local activists and 150 local groups across the country, securing close to 200,000 pledges by September.[13] It also spawned a series of copycat campaigns across Europe.

The impact of Don't Pay reached well beyond the organised resistance. The knowledge alone that people were collectively refusing to pay generated widespread opposition to bill increases. By September, independent polling of energy consumers estimated that 3 million people would stop paying their bills on 1 October, more than half of whom had been inspired to do so by the Don't Pay campaign.

In reaction, the British government announced a £150 billion package to partially freeze energy bills between 2022 and 2024. Right across Europe, similar protests and campaigns took place, with governments instituting a series of price controls and caps in response to the threat of social unrest.

Though not a full retreat by the government or energy companies, the concessions were enough to sap the momentum of Don't Pay.[14] At the time of the package announcement, Don't Pay had just begun to scale up substantially. Some 31,000 people had signed up to become community activists, which, when combined with the existing groups and social media networks, presented an opportunity to build the local defence groups and solidarity networks that would enable participants to resist the bailiffs and forced installations of pre-paid energy meters. Yet despite the fact that the opportunity wasn't taken, Don't Pay stands out as a clear example of how to organise refusal of the squeeze and impose our own price controls.

Anatomy of Refusal

Shoplifting, rent strikes, campaigns of non-payment or fare evasion: all of these involve different forms of organisation and mobilisation. There is no one format to these campaigns. However, we can map the anatomy of refusal as a means of instituting price controls from below.

Against the organisation of our poverty
All successful campaigns of refusal start with our immediate needs, standing against the organisation of our poverty through the price of necessities. We've got to eat, heat our homes, have somewhere to live. We need to get to work or school. When the prices of any of these rise the rest of our

income is squeezed, and the small amount of space we kept for something more than mere survival is sacrificed. The quality of our life degrades.

By focusing on need, the illusion of choice is laid bare. By articulating the squeeze on our lives as something shared, something public, private suffering is transformed into a political terrain on which to organise.

Refusal in this articulation is a social act. It is not a form of passive resistance, a withdrawal from a situation or space. To refuse is to publicly resist. It is an action seen by others. It searches out others who would also refuse, looking for its community. To refuse might seem like an act of negation, but it is profoundly socially productive: it makes community.

During the COVID lockdowns, when many workers were furloughed and forced to subsist on reduced incomes or government support, paying rent became difficult. Eleven per cent of UK tenants fell into arrears in 2020, and upwards of 17 per cent in the US. In response to the rise in rental arrears, a number of large rent strike campaigns emerged, many taking their cue from the Can't Pay, Won't Pay campaign in the US. That campaign called for tenants to stop paying rent from 1 May 2020, in order to put pressure on state governments to implement a series of rent freezes and cancel all rent collection until the end of the COVID public health emergency.

The Can't Pay, Won't Pay campaign worked similarly to the Don't Pay campaign in Britain, through a series of regional and city-based organisers and a pledge system – a mass mobilisation format that uses pledges to act at a specific moment to create leverage. Thousands of Americans pledged to withhold rent, with 14,000 refusing in New York

City alone. In Britain, building on years of student housing activism and rent strikes on campuses, more than 15,000 students from fifty universities embarked on the largest single student rent strike in history.

Each university was organised as a single rent strike, with its own organising committee. Organising committees would focus on collecting rent strike pledges within the accommodation halls of each campus – most British and US universities have large accommodation blocks, with a high percentage of first-year students living on campus in halls. This created a defined campus community. When in Britain the government responded to rising COVID infection rates with a series of mandated lockdowns, students spent even more time in halls and on campus, making organising even more straightforward. Coupled with poor living conditions and remote learning, most students felt they were being cheated by their universities. Through hall-based organising and social-media work, rent strikes quickly gained momentum. As well as withholding rent, students occupied buildings and disrupted as much of the university workings as they could.

At a crucial moment in the strikes, Manchester University students woke up one morning to find the university had erected fencing around their accommodation to stop people from leaving. By 4 p.m. that day, students had had enough. Eight hundred of them streamed out of halls and tore down the fence, inaugurating a month of direct action and escalating confrontational tactics.[15] By the end of the strike, Manchester University had agreed to a 30 per cent reduction in student rents. There were wins across the education sector, though they differed greatly from one campus to another.

As with Don't Pay, the rent strikes managed to reframe the individual act of non-payment as an expression of a mass refusal – a collective act of non-compliance – and a significant number of the rent strikes explicitly focused on forcing government to institute controls or legislative changes to resolve the payment crisis. Many also directly targeted landlords. The leap from what was effectively a protest to a strike – from a campaign to pressure government to one that directly engaged with landlords – was only achieved where there was sufficient organisational depth to produce a physical defence of the rent strikes.

Who sets the price
If the price of necessities is understood to be something beyond anyone's control, there is no point in contesting it directly. At best what can be hoped for is some kind of government help or charity. On the other hand, refusal to accept the naturalisation of the market, or the idea that climate change necessarily makes 'us' poorer, creates new possibilities. Where the actions we could take and the things we could hope for are squeezed, shrinking our horizons, refusal opens up the future. At its core, a refusal denies the limits imposed on us, the efforts to shape and control our actions. It is obvious that refusal in the workplace, epitomised by the strike, is the basis for contesting capital's command over our labour. Just as the wage commands our labour, the price controls our everyday lives.

We can't contest either labour or life without politicising this command. Naked economic interest is not enough. Methods for politicising the price differ, but one of the most common approaches is to refuse the abstractions of the market or government budgets and, instead, to locate the institution responsible for setting the price and make them the villain.

Every campaign needs a villain. A target is not enough. We need an opposing agent, someone responsible, who can be forced to change the situation. Without a villain campaigns can, at best, be little more than protests calling for government aid. In the absence of a villain who can be made responsible, a surrogate is called for. In most instances that surrogate is an actor who can dictate or set the price, despite not being the actual vendor. This could be a company, or regulatory institute, or government body. Lack of direct financial exposure to consumers makes this a much weaker target, harder to pressure into action, and often falling short of the power of the original vendor. But sometimes this is the only target available.

How they come to function as the villain is crucial, as this is how the price becomes politicised. Are they a land-lord who rents properties in terrible conditions, making vulnerable tenants ill while gaining huge profits? Are they a power company charging astronomical rates while raking in hidden fees and charges? Are they a publicly owned bus company whose ticket prices leave people stranded and unable to pay?

At each moment we need to trace the price back to a villain we can confront collectively. Too small, and it is not possible to assemble a collective refusal. Too big, and the villain risks become too distant and abstract for our purpose. To be successful the campaign must find the point at which the market coheres, while identifying how the villain monetises our suffering.

In the case of Don't Pay, the story of record utility-company profits set against astronomical price rises made it easy to find a villain. Journalists had already done much of the early work of research into British utility companies, with scandalous bonus payments for CEOs and record profits, both making headlines. Further research demonstrated that companies were hoarding billions and reaping super profits at consumers' expense, which made for compelling campaign media.

The work of discovering the villain is also the work of finding the community of refusal. Campaigns of refusal often start with a simple call-out or meeting of people affected in the same way by the impossibility of meeting their needs. The initial work of this group is to articulate how they are suffering and to find the villain; and it is the villain that determines how the call to refuse will be structured, and who can respond.

Naming our antagonist, denouncing them publicly, refusing to comply with their demands: this enables us to overcome our isolation as consumers and the obstacles to our collective organisation thrown up by the alienation of the price.

Refusals and demands

The core of the campaign and organising effort is the refusal to pay. Refusals are not demands – you aren't asking for change or permission. Refusal is the grounds on which demands can be built, producing socio-economic disruption instead of mere advocacy campaigns.

Refusal is an active denunciation; a withdrawal of consent. It is an act of sabotage against the price mechanism. The normal functioning of the market is rarely perceived as a system of control. People rarely understand themselves as coerced, because they feel they are making choices that are rational, beneficial or normal. But when prices surge, and supplies are disrupted, the manufacturing of consent falters. When the promise embedded in consumption is broken, the system starts to break down.

In these moments, the transformation of desperate acts of refusal into organised resistance hinges not on rage but on our commitment to each other.

To refuse is hard. Costs and consequences are both high. Organisationally, the temptation is to begin with simpler demands, thinking you can build towards more militant acts of refusal. But this risks starting off with weaker bonds and a shallower mobilisation. Refusals are built on commitment, and so while the appeal might not be as broad, the relationships within the community of the campaign will be stronger.

Campaigns of refusal proceed in the opposite direction to workplace organising or political campaigns. Workplace organising proceeds in steps towards the strike, starting with small acts and escalating towards open conflict. This builds a depth of collective power within the union. Political campaigns are organised around demands that function as landmarks, where the demand is either where the campaign would end (no more arms sales to Israel) or a way to orientate a movement (stop the deportations). Most demands from formal political actors and organisations function in the latter sense, as directional demands.[16] While each demand is an end in itself (such as the four-day week), the idea behind them is to steadily build the world we want to achieve through incremental changes or 'radical reforms', which together realise a broader socio-economic and political programme.

In contrast to these, campaigns of refusal rapidly move towards a political confrontation. This is as it should be.

From the foundational work of first finding community and organising the base, the immediate next step is to set up the confrontation with the villain, be it a landlord, a local council, a gas company or a bus operator. The confrontation might be initiated through demands, but these are always built on the reality of refusal:

'If the fare increases, the city will stop.'

In 2013, a series of Brazilian municipal governments announced fare increases for public transport. Building on previous campaigns against such increases, the Movimento Passe Livre (Free Fare Movement) organised demonstrations in São Paulo. The demonstrations started, much like the Gilets Jaunes and anti-mine protests, with the blockading of roads. The movement occupied major city avenues every second day at rush hour, aiming to multiply support through a campaign that took the entire city as its terrain, and the refusal of all movement as its intervention. The protests quickly snowballed; by day five, 100,000 people had taken to the streets. A few days later, close to 2 million people across 120 cities had joined. Despite severe police repression, the movement forced most municipalities to rescind the fare hikes, and what had started as a revolt against fare increases became a much broader insurgency against the lack of social services, high food price inflation and high taxes.

The reality of confrontation

Refusal inevitably leads to confrontation. It is the insistence on the possible against the structures that manufacture consent and delimit the scope of our imagination and action.

Acts of refusal will be met with violence: the violence of seizure or arrest, the impounding of vehicles or eviction from housing. Cops and bailiffs at the door and court

summons through the letter box. Refusal takes direct aim
not only at revenue and profit, but at the right of capital to
command and government to legislate. It contests power
directly, drawing it into confrontation. And it is precisely
this kind of counterpower we need to build and wield if we
are to combat the squeeze and ensure we are not the ones to
pay for the transition.

Establishing strong networks that undertake both mutual
aid and community defence, therefore, is critical to the act of
mass refusal. Exemplary here is the British experience of
resisting the Poll Tax. In 1989, Prime Minister Margaret
Thatcher attempted to impose a regressive household or 'poll'
tax. This prompted a sustained and widespread campaign of
opposition and non-payment. It proved to be one of the largest and most militant anti-tax campaigns in British history,
one whose culmination famously saw pitched battles between
protesters and police on the streets of London.

In response to the poll tax, local groups known as Anti-
Poll Tax Unions (APTUs) were set up across Britain, and
one year into the campaign of resistance over 1,000 of these
had been established. Initially organized by local activists,
APTUs quickly grew to include diverse members from
across political backgrounds, united by a shared sense of
injustice and economic burden. The unions were primarily
composed of regular, working-class people who viewed
non-payment as a necessary form of resistance, particularly
given the recent memory of the harsh governmental response
to the miners' strike. Unlike mainstream labour unions,
which called for protests and sought a reversal of the tax,
APTUs advocated for, and crucially organised, direct action
and non-payment, since many low-income families simply
could not afford to comply with the new rates.

The APTUs conducted various activities typical of campaign groups, such as disseminating information, organising meetings and holding protests. However, their defining feature was a strong programme of community defence, offering legal and practical support to those refusing to pay. As government efforts intensified to crack down on non-payers, APTUs mobilised entire neighbourhoods to resist attempts by bailiffs to repossess property. By coordinating community efforts to block bailiff access, APTUs became critical in sustaining the non-payment campaign and fostering a shared resilience among working-class communities, ultimately posing a significant challenge to the Poll Tax's enforcement.

It only took five or six people, often a group of friends or people who had been on other campaigns together, to set up the local unions that were the backbone of the anti-Poll Tax campaign. These local organisations cohered opposition by physically bringing people together: meetings were in person, and recruitment worked by either going door-to-door or through public assemblies where people could talk to each other and see the strength of feeling for themselves. The same idea underpinned feminist consciousness-raising groups in the 1960s and early 1970s, where women came together to discuss their everyday lives and struggles in and against their gender, connecting with others who had similar experiences, revealing their struggles to be part of a broader pattern of oppression.

A strong base built on solidarity and mutual aid is essential to carry the refusal onto the two key sites of confrontation: the legal/formal terrain of courts and government bureaucracies, and where the price intersects with our lives, be it the home or the shop.

In most instances, the formal – legal – arena is the first terrain of contestation. Businesses and government departments will often start with intimidation tactics – threatening letters or notices of prosecution. The work of collective refusal starts with knowledge – of legal rights, of what might or will happen – and often one of the main activities of any campaign is the work of informing participants of their rights and the processes involved. From there the aim is to turn the formal processes against themselves, using the rules where possible to undermine the collection of personal data that would enable enforcement. From the mass refusal to register or volunteer information to the occupation of offices and the disruption of data processing, through to the jamming up of courts through mass attendance and disruptive tactics: the goal is to not only obstruct the formal processes involved but to make enforcement difficult, if not impossible.

One of the most successful tools used by anti-poll tax organisations was to encourage widespread non-compliance – refusing to register for the tax, using local courts to contest council orders, and refusing to attend court to slow down the process. The campaign's approach to court proceedings is a useful guide to how this refusal worked in practice. People would be encouraged to turn up at court for minor infractions, as this makes the mass processing of cases impossible, holding up the prosecution. People would make every effort to delay and disrupt proceedings, meaning fewer people could be prosecuted. The gallery would be packed, and crowds would gather outside. More than once courts were occupied, making prosecution impossible. Prosecutions advanced so slowly that it became obvious that very few people would be prosecuted, as it was a physical impossibility. By late 1990, 20 million people were facing court summons for non-payment. The refusal to pay created a financial crisis for local governments, who were presented with the near impossibility of prosecuting non-payers and collecting the tax.

The scale of the opposition culminated in the infamous Poll Tax Riot in March 1990, when over 200,000 protesters gathered in London's Trafalgar Square, leading to violent clashes with the police. The riots, combined with the large-scale non-payment campaign organised by the APTUs, created a crisis of legitimacy for Thatcher's government as public resistance revealed deep divisions within the ruling Conservative Party. The Poll Tax debacle weakened Thatcher's authority and eroded her support among MPs. By November 1990, Thatcher faced a leadership challenge within her party, ultimately resigning.

The relentless opposition, protests and mass non-payment weakened the government's ability to enforce the tax, effectively turning refusal into a war of attrition. In a war of attrition, the objective is not to seize control of key sites or institutions, nor to overwhelm the enemy with force, but to wear down your opponent, making it impossible for them to keep fighting. By turning the lawfare of government prosecution and debt-collection into a debilitating grind, all while draining the resources of both police and bailiff companies, the anti-Poll Tax campaign sapped the government's ability to impose the new tax.

Successful refusals throughout the transition will most likely lead to arrest or the seizure of goods and money. As with recent actions to block immigration raids and deportations, only the organisation of physical resistance is able to halt this repression. Numbers alone are not enough. If you are stealing from a supermarket, you want to not only be too numerous to be stopped, but also able to prevent security guards from detaining anyone. If bailiffs come to install energy meters that force you to pre-pay for your electricity, you want people mobilised and ready to physically stop them; if they are there to seize goods to cover your debts, you want a crowd able to physically see them off. If the police arrive at your door, you want a crowd able to stop them from detaining you.

Proximity is crucial for our ability to collectively refuse. It is not enough to suffer under the same regime, shop at the same store or have the same landlord – we need to be and feel close. While we can feel proximate through social media, and connect our lives and coordinate our actions, there is a bodily geography to the physical act of mass refusal that can only partially be overcome by using technology. We need to physically come together in order to make refusal something more than symbolic. A geographically diffuse refusal often lacks the ability to cohere into an effective physical resistance.

Physically blocking enforcement can mean more than blocking bailiffs and cops; it can mean everything from destroying turnstiles to prevent the collection of public transport fares, to appealing to the solidarity of unions so their members will refuse to cut off power or process fines.

Refusal here extends well beyond contesting their control. It draws us together, establishing us as an organised crowd, as agents capable of taking hold of the squeeze and reshaping the transition.

Refusal and Hope

Refusal is a starting point, one that makes hope possible at a time when it is all too easy to have none.

The transition is a vast attempt at restructuring our lives, that aims to remake our expectations for what kinds of lives we can expect to live and what we can hope for. While it is framed as a response to the mounting disasters that accompany escalating climate change, the reality is that the transition is about reshaping our consumption in a way that is profitable for capital while seeking to limit the demands we make on our governments. And so the myth of choice is mobilised, to make it appear as though the squeeze is our fault; that the problem is our greed, prima facie declaring any more demands to be excessive or a dire threat to the planet.

The squeeze works to diminish the future. As the material conditions of our lives become disrupted, difficult and expensive, as the promise of the future becomes smaller, the space for politics as the practice of changing what is possible shrinks also. But it only does so when those material conditions are understood to be immutable, and when it's obvious to everyone that necessity is always trumped by the obligation to be realistic. Refusal as a collective act insists on the necessary over the realistic, the possible over the inevitable.

However, refusal is not a singular act. Crowds contesting energy bills collide with petty landlords campaigning against heat pumps; both consumers and farmers demand support, while shoplifters and small businesses directly confront one another. Without organising efforts, this diffuse sense of resistance risks coalescing into a reactionary formation under the leadership of small business owners and regional property magnates.

The landscape of refusal must be polarised. It is clear that any political approach which treats all refusals as the same can only serve conservative, pro-business interests. We must distinguish between those refusals that seek to protect profit margins from those that seek to defend living standards and our collective social reproduction.

Refusing the price we're given takes aim at the sanctity of private property. Collective acts of refusal work so that the tenant sets the rent, not the landlord; the customer sets the price, not the shopkeeper. Refusal dismisses the 'natural' functioning of the market to reveal the politics of price-setting, while contesting the right of business and government to do so against our interests. Refusal remakes the market in political and moral terms.

The refusal to pay is a stubborn insistence on the idea that the price is a political object. This marks a break with decades of neoliberal orthodoxy, where economics was insulated from social and political pressure to shore up faltering profits and restore ruling-class power. Contesting prices from below also pushes back against the interventions of the transition economy, that put a floor under prices in order to ensure 'reasonable' returns for investors and green industries.

There has never been a better time for refusal. As neoliberal orthodoxy unravels, and as we lurch towards a new kind of green capitalism in fits and starts, space has opened up around the nature of markets, what can be done, and by whom. Price controls do not seem so far-fetched; at least, no more so than any other kind of government intervention.

Refusal unlocks possibilities. Faced with the contraction of hope through the climate squeeze, our collective refusal to pay the price, to have our lives reduced and horizons shrunk, does not only create new possibilities, it insists on their realisation. But refusal is not enough. To make good on the promises created by our collective refusals, we need to do more than just refuse their transition – we need to make our own.

6

Community Transitions and Counter-planning

No matter how hard we fight, we need to be honest that jobs will be lost and communities dramatically reshaped as industries are transformed or closed down. While car manufacturing and steel are the industrial frontlines now, it won't be long until more industries are affected by transition restructuring. Climate impacts will erode any sense of economic stability and job security, deepening our sense of financial anxiety.

Yet work isn't the victim of the transition – the transition is work. The transition economy is made not only through the 'heroically masculine' work of manufacturing and construction, but through all work, waged and unwaged. This makes work a critical terrain of struggle in the war of transition. Not in order to defend the jobs we currently do, but to tackle the possibility and necessity of transforming work itself.

Beyond Defensive Struggles

We can't fight the war of transition on the defensive. Nor can we fight it one workplace or industry at a time. The zero-sum nature of the transition economy makes it all too easy for capital to relocate industries somewhere less threatening to profit margins, or to withhold investment completely. We can't rely on all industries to continue to provide as many jobs as they have in the past, or even to exist at all.

Some on the Left hope to find a single workforce that can leverage its power against the whole transition: an Archimedean point against capital. Power and manufacturing are the usual candidates on which hope alights when seeking another industrial vanguard. But despite scouring new frontier industries, no strategic sector has been found. Nor is one likely to emerge.

There is no central front – no single point of leverage. Not only is there no one single place to focus our organising efforts or political hopes, work itself has shed much of its promise of class unity. For most of us, work is something we suffer through – a source of anxiety, a place of bullying or abuse, or an activity we find devoid of substance or meaning. Often it is a place where any sense of investment in work is betrayed through the imposition of ever more metrics and bureaucracy and the steady erosion of our autonomy and working conditions. Most jobs might not be 'bullshit' (though clearly many are), yet work decreasingly holds any promise of satisfaction, and increasingly seems like something to minimise in our lives as much as we can.

Community Transitions

While no single workplace or even sector may constitute a point of systemic leverage, this doesn't mean we can put organising workplaces to one side as a tactic in the broader struggle for a different transition, one that meets our needs at home and at work. Workplace organising in the transition economy cannot be focused on workplace issues alone, however, especially as many of the reasons given for changes made in the name of the transition exceed the limits of any one factory or office. Nor can we argue against restructuring and closures without reference to life outside the factory gates. Organising within and against their transition means starting with a 'whole-worker' approach.[1]

Whole worker organising is a labour organising strategy that views workers not just as employees with narrowly defined workplace concerns, but instead as people embedded in their communities that face issues beyond wages and working conditions. This approach seeks to mobilise workers by addressing the full spectrum of their lives, including their roles as parents, community members and citizens, in addition to their roles as employees. As such it is built on deep webs of social ties and rests on the assumption that workers have strong relationships across communities that can be both organised and mobilised to support workplace struggles.

This approach is in line with how collective politics within the working class was historically forged. The consciousness of a class 'for itself', of a working-class collective identity and set of interests was built not solely, or even primarily, through workplace organising but through social and sports clubs, community resources and ventures, social services, cultural events and working-class media. 'The community' was once a key organising terrain, where working-class power was organised and exercised.

Whole-worker organising starts from the premise that to effectively organise collective power within a workplace, you have to organise not just the workers but the broader community they are embedded in. For some industries, such as steel production, it's clear that the call for community support stems from the economic impact of lost jobs. For other industries, particularly those that produce not only profits but social value, such as hospitals and schools – industries of social reproduction – community engagement extends far beyond the economic.

For these sectors, and many other public and service sectors, the community is more than just dependent on the employment these offer – the community is the site of production, with citizens or consumers being intimately involved in the production process itself. This intimacy enables clients and consumers to see and engage with workplace conditions, and to feel a sense of ownership over the work. We need to understand the community as both product and part of the production process. When the community is the site of production, what is called for is not whole-worker organising, but whole-community organising.

As well as a raft of changes to industry and employment, the transition economy involves the remaking of our lives. It is a profound reorganisation of community, undertaken through the workplace. For workplace organising in the transition to be successful, it needs to be whole-community organising. Starting from the community rather than the workplace enables us to come at the war of transition from another angle, one than opens up the political space of the transition and moves us past simply counting carbon emissions within a business framework.

It also pushes us to think about what we need in a collective sense, forcing us to confront the main question: if not their plan for the transition, then what?

Back to Port Talbot

We can see why we need to shift to a whole-community organising perspective more clearly if we return to the steel mill in Port Talbot. We left the workers fighting for their jobs, in the face of the mill owners looking to shut down the highly polluting blast furnaces and, with massive government support, switch to using electric arc furnaces. While this would eventually lead to huge reductions in carbon emissions, it would also cause at least 2,800 job losses (out of 4,000 workers). Should no other major industry invest in the area, the impact on the town of Port Talbot would be devastating, with five to six times as many additional jobs lost in the local economy because of the closures. As has happened all too many times before, deindustrialisation would likely lead to profound social devastation, this time justified by the imperative of reducing carbon emissions.

The business plan for Port Talbot is to demand massive government subsidies – in line with most other green industries, such as electric vehicles – and then focus on technological substitutions to realise reductions in carbon emissions. Direct subsidies, coupled to efficiency and wage savings will ensure profitability, while the government is able to boast that it has both secured some future for

British steel and made progress on meeting its legally bind-
ing climate targets.

The unions' counterplans are likewise focused on mobilising government support and new technologies, motivated by
the need to save existing jobs rather than Tata Steel's profits.
It's useful to look at those counterplans in order to illustrate
why a single-workplace approach will not be effective, especially in the context of the zero-sum green transition.

The most comprehensive plan was put forward by Unite
the Union. It essentially boiled down to keeping the blast
furnaces running until the end of their functional lives
(2027 and 2034), while installing an arc furnace alongside
them, increasing overall production. It further suggested
looking to modify the blast furnaces to use hydrogen, as is
being trialled in other countries. Unite also called for deep
investment by the government in the entire region to turn it
into an industrial hub, specialising in green industries such
as wind turbine production.

Delaying the closures of the blast furnaces would mean
delaying the reduction of the mill's carbon emissions. The
British government is heavily investing in hydrogen development, alongside massive private investment, but there is no
definite commitment to any one steelmaking project as yet.
Hydrogen – specifically green hydrogen – has been put forward as a future wonder-technology, especially for heavy
industry. Across Europe, billions are being invested by governments and the private sector in developing hydrogen
steelmaking. In order to be 'green', hydrogen needs to be
made with renewable energy, which consumes a huge amount
of energy. If British steel mills were to use hydrogen in the
production process, it would require sixty-three times more
offshore wind-power capacity than the combination of what

currently exists and is planned. The switch to hydrogen would also entail job losses, as the process is less labour-intensive than fossil fuels. Finally, as we've seen, there is little hope for a British manufacturing revival, or one that would generate large numbers of jobs – not without industries surviving largely through government subsidies, and a significant period of deglobalisation.

In the end, though, should any of the above result in making Port Talbot's operation less profitable than it is, or 'scare off investment', Tata Steel will likely just close the mill. Similar problems are faced not only by other steel mills in Britain, but by most of what are called 'foundational industries', such as steel, concrete and glass. All will ultimately depend on government support and novel technologies to remain operational, often with reductions in the total labour force. And opposition plans usually amount to demands that jobs aren't lost, relying on either expanded production or, similarly, novel technologies. Clearly this is a zero-sum game – you cannot increase production and job numbers everywhere in a global market.

In brutal terms we know that change comes when bosses or politicians are forced to make concessions or agree to changes because workers (or citizens and consumers) inflict 'economic distress'. A more expansive way of understanding this is that power comes from the organised ability to contest control over production – how it happens, under what conditions or whether it happens at all.

There is a basic math that underpins victory in this approach. Bosses and politicians weigh the cost of giving in to worker (citizen) demands – the concession cost – against the cost of the disruption.[2] If the concession cost is lower, they are much more likely to concede. The more disruption you can inflict, the more power you have. In other words, if you can shut down a profitable business, and the owner has no other business or investment, then you have much more power than if you shut down a business that is unprofitable but just one operation in a large corporate portfolio. As in the case of Port Talbot.

The fundamental problem facing industry during the transition is that it is being transformed by technologies that require fewer workers, in a context of intense international competition and economic stagnation. No single workplace, or even national sector, has enough marketplace leverage in such a situation. Of course, this leaves aside the question of whether we should collectively want Port Talbot to keep its blast furnaces open – in its own small way aggravating climate change for all of us.

From the perspective of looming ecological collapse, losing just under 3,000 jobs is a relatively small price to pay for cutting 1.5 per cent of Britain's carbon emissions. Here is the sort of budgetary logic that too often characterises not only government policy but contemporary climate politics. But those 3,000 jobs, and the many thousands more in Port Talbot, are not expendable. Any more than are the people living in the vicinity of planned wind farms or lithium mines, sacrificed in the name of a low-carbon future.

The difference in vision between Tata Steel and Unite boils down to the matter of work. Tata Steel sees a future with less work in it, Unite sees one with more. Given the devastations of decades of deindustrialisation and government austerity programmes, it's not surprising that work has become the central ground of hope for the future. It's also not surprising given that steel workers often report higher levels of job satisfaction than many other workers (along with higher wages, which might have something to do with it).

But when canvassing the Port Talbot community, the main reason people give for supporting strike action isn't that the jobs in themselves are satisfying, but that they fear what closures will do to the community. While bosses and unions fight over jobs, the broader fear is for the existence of the community itself.

The overriding concern for the workers at Port Talbot is how they'll pay their bills once the jobs are gone; for the community, it's what will happen to the town when the people who live there have been impoverished; for all of us, the bigger question is how to ensure a habitable world within which we can survive. At first glance this would seem to be a classic example of jobs vs the environment, but, as we've seen, a defensive battle is unlikely to result in a consolidated long-term victory. So how do we square the circle?

Within the transition, what is under attack is the future itself. The future we had is lost, and defensive tactics not only won't bring it back, but will also isolate our struggles. The fight in Port Talbot isn't one of jobs vs the environment – it's against the destruction of the future, a destruction organised through their transition policies and plans.

But what happens if we make the future the basis of our battle, and contest the direction of their transition? What if we go beyond defensive workplace and community organising, and put our plans for the future we want at the heart of our struggle?

Lessons from Radical Municipalism

If we started from what we want at a community level, it would likely look similar to the wave of radical municipal struggles that emerged after the 2008 financial crisis.

Broadly speaking, municipalism as a political framework encompasses both the practice of self-governance by towns, cities and city-regions of various sizes and the viewpoint that supports these forms of local governance. On its own, municipalism appears politically neutral – there is nothing inherently radical about it as an approach. A radical municipal strategy views the municipal level – including how people's lives are structured in these areas and the institutions that govern them – as a site of political contestation. Rather than seeing municipalities as depoliticised, lower-tier administrative units under the nation-state, with limited political significance, a radical municipalist perspective considers whether organising at the municipal level offers a distinct revolutionary potential.

There is no one model of radical municipalism, but all pursue similar goals. They all look to take control over economic processes within a city or region and subject them to community direction; they politicise economic decision-making and planning. They all try to maximise the tools of local governance, including procurement, public services and regulation, to create a democratised economy that raises wages and improves working conditions. And they all attempt to deepen the bonds of social solidarity and inclusion.

Some of the most notable examples include Barcelona en Comú, a citizen-led municipalist platform founded in 2014 in the Catalan capital, which emerged from Spain's anti-austerity movement after 2008, and Cooperation Jackson in the United States. Cooperation Jackson was also founded in 2014, with a vision of building a network of cooperatives to promote economic democracy in Jackson, Mississippi, rooted in Black liberation and radical municipalist principles.

Perhaps two of the most significant experiments in this vein come from outside of the Global North. The Zapatistas have long pursued radical muncipalist approaches to the articulation of indigenous autonomy in southern Mexico, through the Municipios Autónomos Rebeldes Zapatistas. Until 2023 these were the basic municipal units in Zapatista territory, and power rested with the popular assemblies in which anyone over the age of twelve could participate in decision-making. Elsewhere, the most profound example of radical municipalism in practice is the Autonomous Administration of North and East Syria (often referred to as Rojava), which has successfully implemented a system of directly democratic assemblies involving millions of people.

One important development within radical municipal frameworks was the idea, developed in Preston in north-east England, of building a radical municipal project through what were called 'anchor institutions'. The 'Preston model' targets significant economic institutions, specifically large local employers who are unlikely to relocate and who command significant procurement budgets, making them powerful economic multipliers. One of these is local government, but they also include hospitals, universities and schools, utilities and housing associations and transport services. The approach to these institutions is twofold. The first is to advocate for better wages (living wages) and conditions across these institutions, in order to drive down poverty and increase equality (and indirectly boost local spending). The second is to convince them to direct their spending and procurement towards local businesses, preferably worker-led cooperatives, social enterprises and small and medium-sized businesses.

The second element of this approach reflects the emphasis on public and community ownership and economic democracy. The model promotes the creation and support of worker-owned cooperatives as part of a broader strategy for building community wealth. At the same time the model calls for the development of community banks and credit unions, as well as the municipal ownership of key assets and infrastructure, such as public utilities, housing and energy projects, the latter of which could be powerfully mobilised as part of a democratic energy transition.

The Preston model has secured a series of important gains, including a marked increase in public-sector spending going to local suppliers, credited with revitalising the local economy, reducing unemployment, fostering cooperative businesses and promoting fair wages, and is seen as an important model for other cities and regions. It has significant limitations, however.

As with other radical municipal projects, it is constrained by national legislation and regulations, and by cuts to council budgets. So far the gains have been on a modest scale, with economic growth confined to specific areas and sectors (although, if networked with other municipalities, this could be overcome somewhat). Most importantly, it is a consensus-driven model, not a conflictual one. Anchor institutions are asked to change their behaviour, not forced to.

Here we can see the advantage of bringing municipal approaches together with whole-community organising. Where municipal approaches enable economic and community planning at scale, along with a broader emphasis on community needs and institutions, workplace- and community-organising strategies build disruptive forms of power that enable the implementation of community counterplans. It is the building of community- and workplace-based power that facilitates the move from asking anchor institutions and key economic actors to participate in the municipal plan to forcing them to do so if they resist. Additionally, working at a municipal scale permits gains to be embedded in local governmental institutions, making them more robust and enduring.

A municipal approach also enables the parallel creation of economic policies that foster a democratisation of the local economy and the creation of new anchor institutions, in particular providers of housing and energy, the latter vital for the democratisation of the energy transition. Local financial institutions such as credit unions and community-owned banks further develop this economic democracy.

The combination of municipal approaches and whole-community organising would impact a sizeable proportion of local employment and economic activity. Even more importantly, coordinated at a municipal level it would create the possibility of economic and social planning at scale. By bringing planning into our organising efforts, we will be able to push past our defensive campaigns and build towards a more radical horizon. Planning is essential to redirect the transition, and whole-community organising builds a depth of power that enables us as a movement to go beyond the sphere of climate policy.

Counterplans and Community Transitions

'When we were hired, nobody asked us what we wanted to produce. So we don't accept being blamed for what we produce when someone fires us. If we could choose what to produce, of course we [would have] other ideas.'

> Dario Salvetti, Collettivo di Fabbrica
> (GKN Factory Collective)[3]

The mass sacking took place via email on a day off. On 9 July 2021, all 422 employees and around eighty temporary workers at the GKN automotive factory in Florence, Italy, received termination notices without any previous indication that the future of the plant was at risk. The closure was blamed by the bosses on the transition to electric vehicles.

In response, workers gathered at the factory gates to protest. They pushed past the security guards deployed to block the gates and occupied the building. Using a tactic legally codified in Italy, that of forming a permanent assembly in the factory, they declared they would keep the machines running and the factory open. The union subsequently obtained a court ruling that the sackings were illegal, and the workers are now in the process of trying to secure funding to take over the ownership of the factory.

Such occupations in response to closures are not uncommon. What caught the imagination of so many, however, was what GKN workers did with their occupation. Since the factory closure was due to the pressures of the green transition, workers proposed to lean further into transition economics: they put forward their own counterplan as to what the factory could produce. At first workers decided in assembly that they could produce parts for public buses, but the local government was not supportive (imagine if the council was controlled by pro-community transition politicians!). Workers then developed a plan to move from automotive parts to cargo bikes and solar panels. In both instances, workers decided that in response to the transition, rather than fight a defensive battle they should seek to turn their productive capacities towards the transition.

This counter-transition is still in progress. The factory produces cargo bikes, working with local farmers to create an agricultural cooperative that uses the bikes as a part of the supply chain. They are also attempting to secure further investment in cargo-bike production, as well as more for solar panel production (again, imagine if there was a community bank that could directly invest in the workers' cooperative). There is currently no consistent income for the 120 remaining workers; most of the rest had to leave to find other jobs. Despite strong local support and a real need for both products, the future is still uncertain.

There is more than an echo of the Lucas Aerospace struggle in the GKN occupation. Lucas Industries was a Birmingham-based British manufacturer of motor industry and aerospace industry components.[4] In 1974 Lucas Aerospace announced job cuts and restructuring. Half of its production focused on fulfilling military contracts and was funded by government spending, and Lucas Aerospace was struggling to compete with emerging aerospace and defence manufacturing giants in the US and Europe.

In response to the announced job losses and with no help forthcoming from the then Labour government, workers organised. They spent two years developing a plan to repurpose the plant from making weapons to producing socially useful goods, including advanced green technologies like wind turbines and hybrid power packs, and medical equipment such as kidney dialysis machines. This is what became the Lucas Plan, perhaps the most famous example of an ecologically focused workers' counterplan.

The plan attracted broad interest and support, as it proposed not just an alternative to save jobs, but the subordination of production to social use, declaring there to be a right to participate in socially useful production. It was part of a movement that argued that technology must be shaped directly by social choices – that development was something to be contested from below as a democratic process.

When radical Left candidates won power at the Greater London Council (GLC) in 1981, they introduced an industrial strategy explicitly committed to socially useful production. A series of Technology Networks were created – community-based workshops, with tools and technical advice available to all.[5] These spaces aimed to democratise scientific and technical knowledge and production, and hundreds of designs and prototypes, for everything from wind turbines to electric bikes, were produced over their short lifespan. The products and designs were registered in an open-access 'bank', while the Greater London Enterprise Board provided assistance to turn ideas into cooperatives and social enterprises.

For all the thoroughness of their proposals, Lucas Aerospace workers failed to pressure the company into implementing their plan and the proposals were rejected by management. This was despite the fact that analogous (or sometimes identical) work was already being carried out on some of the Lucas Aerospace worksites, and that the company had already recognised that aerospace work was in decline.

Both GKN and Lucas Aerospace provide a model of bottom-up democratic planning, not only of production but, crucially, of innovation and technological development. Both demonstrate that workplace organising, even in the face of mass layoffs and closures, does not have to be limited to the defence of existing jobs but can expand to encompass a more radical agenda.

Community Planning as Political Strategy

Communities are produced. The labour of producing communities is not confined to waged labour or formal workplaces, but takes place, and is woven, throughout our communities.

Social value is not the same thing as economic value. What might be good for business might be destructive of individual lives or community relations. This means we need to approach the production of communities by looking not at strictly delineated areas of activity, but at what work is being performed, regardless of whether it is waged or unwaged, public or private.

If we are to formulate a strategy from below, we need to focus on our needs. Focusing on our needs – on socially useful production – emphasises the centrality of social reproduction to our lives and communities. This acts as a corrective. Whereas focusing on particular workplaces makes the preservation of waged jobs the central consideration, emphasising social reproduction means starting from the totality of our existence and prioritising life over work.

Social reproduction, simply defined, is the creation and maintenance of our lives, both physical and social. Long a staple of Marxist-feminist critique, using social reproduction as a political lens enables us to see that it is our labour that creates and reproduces society as a whole.[6] And while there is a significant amount of waged social reproduction, from education to healthcare, most of the work is unwaged and relies on the invisible labour of (overwhelmingly) women. This labour is not 'accidentally' unwaged – it is crucial for the work of social reproduction to be devalued and as 'free' to capital as possible.[7] As with environmental resources, capitalism relies on human labour being cheap; and it can only stay cheap as long as most of the work required to produce and maintain it is done for free.

Recentring the labour of social reproduction is all the more critical within the transition. As the squeeze takes hold, the demands on social reproduction will only intensify. The accelerated crisis of social reproduction engendered by the climate emergency forms a crucial frontline in the war of transition. It is here, within the crisis of social reproduction, that we find the most fertile grounds for political organising.

Ironically, focusing on the labour of social reproduction also brings the emptiness of much waged labour into sharp relief. If a dwindling number of professionals still invest in their work, drawing meaning from their roles, for most of us work is something to be resented. It is not just that we are compelled to sell our labour to pay the bills. It is also the way so much work is devoid of social value or meaning. Often it is not only bullshit, but destructive.

Despite the huge efforts poured into coercion and control, not to mention the vast sums spent on selling us the importance of a 'work ethic', the widespread and deepening refusal of work – from campaigns for a four-day week to post-COVID phenomena like 'quiet quitting' – amounts to a persistent threat to workplace order and capitalist productivity.

Bringing our desperate personal need to secure waged work into conversation with what we need as a community opens up a space to question the value of work itself. Yet this question, much like that of the vital importance of social reproduction, is unanswerable within existing economic frameworks. The problem is to act on these realisations as part of a practical campaign against the existing transition economy.

Planning, done through whole-community organising, enables us to act on both of these realisations. It is through community deliberation, dialogue and planning that we can move from isolated struggles to the imposition of our counter-transition.

The foundational unit of a community organising campaign is the community assembly. This assembly is the deliberate construction of a coordinating body of union and community activists, and workplace organisers. Like any initial organising body, it does not need to start large, but it does need to prioritise expansion and consolidation. As with a workplace, the first task of the assembly is to recruit more members from as many of the key economic institutions across a community as possible. From there, the next step is to build membership and power in each institution, creating as large and coordinated a disruptive capacity as possible. Ideally the focus is on those economic institutions that are both the most socially productive (care homes, hospitals, public utilities, etc.) and large enough to act as anchor institutions for the whole community. At the same time, organising efforts should be extended beyond formal workplaces – not only into small and medium-sized businesses and contract labour, but into community organisations and the home itself.

As the assembly grows, and as its ability to coordinate disruptive power increases, it will become capable of directly coercing some political and economic actors, including government; it will also be capable of directly taking over others. At some point in its development it will need to secure its gains through local government. While radical municipalism has its limits, there are very real gains to be made at an institutional level.

What's more, underpinned by whole-community organising, taking municipal power paves the way for a series of community-wide interventions, from the use of procurement policies to community finance, through to the promotion of cooperatives and social enterprises, investment strategies and legislation. It is tempting to think that such a programme can best be initiated from the top down, and that it would be faster and somewhat easier to just take state power. However, even at a local level, the past decades of municipal experiments, and the electoral turn of the Left in the Global North after the 2008 financial crisis, demonstrate the limitations of such programmes when they lack a strong social base capable of exercising power against established elites and institutions.

Shifting the focus of planning from a single workplace to what a community needs directly connects the transformation of the workplace to socially useful production. It also throws open the question of what a workplace does to a much broader constituency. Considered as a social resource, workplace planning can then incorporate broader social agendas and visions, from access to employment and training, as well as production. But more than this, it allows us to broaden the emphasis from what we can do now to how we want to live in the future, not as an abstract vision implemented from the top down as some kind of utopian social policy, but as the direct result of our labours.

Whole-community organising and building counter-plans through workplace and community assemblies is not a total solution. But focusing on the medium scale of local communities – more than a workplace, less than the fantasy of seizing government power and being able to smash down on the magic state button – enables us to build a democratic form of power grounded in our everyday lives and accessible to people beyond the wage relation. It forms a base for our struggles. Whenever the working class has exercised real power historically, it has done so not on the basis of a single factory or work site, but on the strength of its communities.

The Work of Adaptation

After yet another 'hottest summer ever' in 2024, Britain's National Education Union (NEU), launched a campaign for a maximum working temperature in schools. Surveying their members, the NEU found that 85 per cent of teachers worked in overheated buildings during the summer, with a third having to work – and students to study – in danger-ously sweltering rooms. Despite this, there is currently no British law governing work in adverse heat.

The union's proposals call for national and local regula-tion around working and studying conditions in high temperatures. They are also pushing to create an expanded role for health and safety reps, adding 'environmental' to their brief, and looking for ways to work with their members and the broader educational community to help schools become both carbon-neutral and better adapted to extreme weather swings.

This comes only a few years after the end of the wave of school strikes organised under the banner of Fridays for Future. Launched in August 2018 by Greta Thunberg, by 2019 the strikes had spread to over 150 countries, with mass protests involving millions of students and supporters. Major global events were staged to coincide with key climate summits, like the UN Climate Action Summit in September 2019. Schools regularly saw mass walkouts in support of climate action by students, often in defiance of punitive responses by school authorities. The NEU in Britain supported the student strikers and defended their right to take action. They have, besides, long called for climate change to be recognised as a social justice issue and lobbied for changes to the school curriculum to reflect the importance of climate change.

While adapting working conditions to meet the demands of changed weather conditions would seem to be too obvious to have to fight for, expecting businesses or government agencies to voluntarily adopt changes that will often cost more and disrupt existing procedures is profoundly naïve, as demonstrated by anti-heat–measure campaigns in the US and foot-dragging on the issue by business and government alike. From the bottom line to bureaucratic inertia, working conditions only ever improve in response to the pressure we generate and the disruptions we cause.

While it is crucial to organise in heavily affected workplaces and industries, which are those that make or break a local economy, we need to be careful to not then ignore the grave deterioration of working conditions the climate crisis is creating. It is this wholesale state of deterioration that forms the basis for the generalisation of class conflict over the transition to a net-zero economy.

High temperatures, the chaos of disrupted services and extreme weather events: all of the soon-to-be-normal conditions of life in the climate squeeze form a common social condition which is immediately political in the context of the transition and climate change. Far more than a focus on specific industries or jobs at risk, addressing the broader deterioration of conditions generates a whole-community approach, since the community is the point of production of most services.

Such an approach also foregrounds the broader neglect of climate adaptation. Climate adaptation is the process of preparing our communities, economies and lives for the impacts of climate change. While most countries have paid lip service to climate adaptation, very few have substantive plans to adapt, and there is a common lack of preparation. Where adaptation plans exist, they are often vague and underplay the challenges faced by people and infrastructure alike, as well as underestimating the impacts of climate change. Few are adequately funded. Adaptation is a reactive and haphazard process across the board, which, in conjunction with reductions in public spending and ongoing programmes of austerity, produces a generalised condition of climate neglect.

As with severe heat, the lack of adequate preparation for extreme weather is laid bare by the impacts on services such as schools and care homes. Community campaigns focused on these anchor sites of social reproduction open up two crucial further questions.

The first urgent question is how to reconfigure social and public services, especially those we rely on most, in light of the squeeze. This is not to call for an endless round of think pieces and policy documents, but to incorporate the adaptive transformation of social and public services into our counter-planning for the transition. The squeeze will lead to greater pressure on services and the emergence of new kinds of needs and demands. We must work through what the climate squeeze means for the services we need and how to deliver them, as well as how we adapt them for a disruptive future.

The second question is who is expected to adapt. Not everyone is equally exposed to the squeeze nor to the impacts of the climate crisis at work. Focusing on who does the work of social reproduction emphasises the uneven vulnerabilities in our communities, those who have fewer capacities and resources available to them to implement their own adaptations. But adaptation is not just exposure – it is also the additional work that results from the squeeze and climate change. If gender relations remain the same, this will mean escalating care duties and ever-lengthening double shifts for most women over the coming decades.

While the first question pushes us to return to the vexed matter of how services are delivered – as things we need, but not in the way they are provided – the second illuminates the private nature of contemporary approaches to care and the domestic burden of climate change. Despite some progress, the bulk of reproductive and domestic unwaged labour is still done by women. As the climate crisis worsens and the squeeze intensifies, the amount of work that needs doing will increase amid deteriorating conditions. Yet in the context of a community transition that takes our collective needs as their starting point, it's time to ask whether this work should continue to be distributed so unevenly, or should even remain behind the closed doors of a family home.

The provision of care, health and education is already a political issue, with the poor conditions and high cost of all three providing fertile ground for popular discontent and revolt. It is not a big step to take the private stresses of the double shift and escalating care responsibilities and make them public through the process of whole-community organising.

In the broader context of counter-planning from the community, turning the home inside out alongside the factory follows both politically and logically. If we can counter-plan from the shop floor, then we can counter-plan from the kitchen.[8] Even asking these questions sends us back to solutions long proposed by feminist campaigners: from collective laundries and community canteens, to nurseries and social facilities attached to workplaces, to the radical democratisation of care and health services.

Foregrounding the conditions of work within our counter-planning and organising practices brings us back to the question of how we want to live. Far from the scramble of individualised responses to the squeeze or the hustle of a brutal neoliberal survivalism, asking this question at a community level lifts us beyond defensive struggles into a space of not just political ambition but hope.

The Political Strike

It's not enough to organise towards our own transition; we have to fight against the transition economy being enacted by the consortium of elite actors, from private corporations and individual billionaires to European and American governments. Their transition is not just a project of imposing new tasks and technologies, or even new forms of management and labour relations. It involves the development of new frontier industries and technologies, many of which are actively destructive and socially toxic.

Opposing the buildout of their transition economy is not just a question of community campaigns and environmental protests, but of rediscovering the political power of labour to contest development.

We are already being told that to stop climate change we need new mines, that vast wind farms are an absolute necessity, that hydrogen and biofuels are indispensable for 'hard to abate' industries. We're told that Big Data and Artificial Intelligence are necessary to solve intractable problems and make our electrified economies 'smart'. Above all, we're assured that together these will lead not only to economic growth, but new jobs and the revival of communities suffering the rot of deindustrialisation.[9]

Some of these projects have little to do with arresting climate change, and much more to do with the promise of profit that flows from what are considered promising green investments. We don't 'need' AI or the proliferation of data centres to solve climate change. We may need some hydrogen, but the extent is questionable, and it's worth asking: who is this 'we', anyway? Who decides that 'we' need to keep extracting natural gas to generate hydrogen, or to direct renewable energy – which could otherwise be used to heat homes or run hospitals – towards the production of hydrogen? What is clear is that if fossil-fuel companies are to protect their investments in gas extraction and distribution, both of which they want to repurpose for hydrogen production and distribution, then 'they' need a hydrogen economy.

Most of the development of these new aspects of the transition economy are not 'necessary' at all. What we actually need is to reduce carbon emissions. What we need is justice. To build a data centre, a nuclear power plant, or a blue hydrogen production plant is not a necessity, but a political choice. We could choose to tackle climate change by reducing energy use, or employing less raw materials. We could build smarter and work with what we have. There is no shortage of alternatives that would leave 'us' richer as communities, using less energy per capita. What is often presented as an imperative of the green transition is no such thing – it is a political choice, made by big business and ruling elites looking to protect their power and wealth.

Often these developments only announce themselves after they've been built. In the water restrictions imposed after a new data centre comes online, stressing the local water network; in the mass marine die-off in a harbour being dredged to clear the way for a new gas pipeline; in lorries and cement trucks driving in convoys through local streets. By the time we reach this point, the only tactic that will work is the blockade. But this is not the only point of intervention. Transition is a work of labour – those tasked with building it can refuse, exercising their disruptive power not only to contest wages and conditions, or the delivery of a service or product, but the construction of a facility in the first place. The strike can be rediscovered as a political weapon. As Jack Mundy, secretary of the New South Wales Builders Labourers' Federation, put it,

> Yes, we want to build. However, we prefer to build urgently required hospitals, schools, other public utilities, high-quality flats, units and houses, provided they are designed with adequate concern for the environment, than to build ugly unimaginative architecturally-bankrupt

blocks of concrete and glass offices . . . Though we want all our members employed, we will not just become robots directed by developer-builders who value the dollar at the expense of the environment.[10]

Indeed, throughout the 1970s that trade union in Sydney, Australia became infamous for mobilising the strike as a weapon against noxious development.[11] In 1971, in response to a call from a local community group opposed to a corporate developer attempting to rezone their local parkland for housing, the Builders Labourers' Federation (BLF) imposed the first 'green ban', designed, as the name suggests, to protect the local environment. The BLF, a key construction industry union, declared that no member would work on any development on the parkland, effectively banning construction on-site. When the developer said they would bring in scab labour instead, BLF members gathered on one of the developer's half-completed building projects elsewhere in the city and proclaimed that the building would remain forever unfinished, as a monument to the lost parkland, if development on the green space ever went ahead. The struggle continued for another twenty years, but was eventually won by local residents. This was the first of over fifty green bans, where the BLF contested not only capital's right to command labour, but its right to decide how society develops. More than this, the BLF also used their organisational density and power to contest other issues, including threatening a work ban at the University of Sydney unless it withdrew a decision to not run a women's studies course, and another at Macquarie University when it expelled a gay student.

The NSW BLF's ability to take political strike action was built on deep organising that took place over more than a decade. It was this process of deep organising, coupled with the importance the union placed on political education, that enabled the BLF to mobilise its members in defence of community actions – often against the immediate economic interests of those same members, who would have materially benefited from new construction projects.

This is not the only example of a trade union using its power towards ecological and social ends. Dock workers have often flexed their muscle to disrupt toxic industries such as the nuclear industry, or to confront racial injustice, as in the South African apartheid regime. Logistics and construction industry workers in particular have long exercised this kind of political clout. And given the centrality of logistics and construction workers to the development of the transition economy, both are well positioned to leverage their power against the net-zero buildout alongside impacted communities.

But more than this, the experience of the BLF demonstrates the crucial role played by political education. Far from accusations of 'trade union consciousness' and the economic limits of self-interest, what it shows is that workplaces and unions are fertile grounds for developing a broader political movement. We can see this in the growing prominence of union members in other struggles and campaigns, or in the fact that just being a member of a union makes you far more likely to hold progressive views and believe in collective solutions.

Most unions currently pour huge sums of money into political funds and campaigns, but only a fraction of this is dedicated to member-focused political education. The vast majority of it pays for what should be considered lobbying efforts with political parties and governments. It is money for access, with a poor track record of success over the past forty years. We need to redirect these funds towards the sorts of education campaigns the BLF undertook to build support for political strikes and work bans, and towards the broader social education programmes and institutions that have underpinned historical socialist movements, from newspapers to social clubs. Unlocking the potential held within existing workplace organisations will enable a wider eco-socialist movement to flourish, giving us the grounds on which to fight and win in the war of transition.

The Reality of a Democratic Transition and Living in a 3°C World

In the end, the transition will be a mess. There are no neat policy prescriptions or political strategies that will spare us the hard and necessary work of organising workplaces and communities. Nor is there any way around the fact that much of what passes for work nowadays is socially and environmentally destructive, so that any just transition that aims to bring about an equitable and sustainable world needs to phase that work out. The idea that there are easy, win-win solutions is a fairytale.

Yet to fear that this makes the work of counter-planning and whole-community organising impossible is to sell ourselves far too short.

Of course, there are powerful structural constraints on both workplace actions and on our ability to collectively organise: anti-union and anti-protest laws, pervasive surveillance and union-busting companies; a lack of time away from work, of resources and of collective institutions; and a lack of social services and decent work, pushing many of us further into a hustle mindset. However, far too much is made of these obstacles. People have long organised in far worse conditions, with even fewer resources and much less time away from work and the demands of just staying alive. At the historical high points of union power or radical social movements, we worked longer hours and faced more brutal repression from governments and capitalists alike. The truth is that we have never made gains from a position of comfort, and there is no hard barrier to us organising now.

It could be argued that focusing on local organising – at the scale of a community, town or city – is far too minor. That this sort of approach to the politics of the transition and climate change risks achieving very little, whereas the magnitude of the present crisis calls for nothing less than seizing the commanding heights of the global economy.

While this argument erroneously confuses the scale of the problem with the unit of social change – because change isn't an abstract, general thing that happens to an entire system; it takes place somewhere – the main problem with analyses that focus on state-level change or general policies is that they lack political substance. Beyond calls for a renewal of union activism, for environmental activists to act differently or for the creation of a new popular political party, it is never made clear how the state will be seized or power exercised. What will unions, or activists, or even political parties do differently? And who will do the work to take up these calls?

Nor do any of these prescriptions recognise the need to question the actual history of trade unionism in the Global North, the limitations of activism as a mode of political action or, more crucially, a hundred years of electoral failure. Any call to seize the state must first set out why it would be different this time, especially given that the call takes place without the backing of a broad and powerful mass socialist movement.

What recent electoral experiences have shown is that without a broad and organised base, one that has not only depth but political autonomy, no electoral strategy has any hope of producing change at all. Focusing on a community transition enables building that base while actively fighting the war of transition.

It could also be argued that we don't have the time to patiently build community and workplace power, that change has to happen faster, because the climate can't wait for us to undertake the hard work of political organising.

We need to accept that this is true but doesn't alter anything. There is no perfect policy or media message, no magic change button to push. Rushing to government on a vague, inclusive win-win platform might, with a huge amount of luck, win votes, but it offers scant hope for actual lasting change. Power comes from organising, and organising takes time and courage.

We need to accept that we are organising within the climate crisis, not against it. We must fight to limit the damage, to keep temperature increases as low as possible – every fraction of a degree counts. But all politics is now a politics of adaptation.

7

Struggling for a Liveable World

Claim no easy victories.

Amílcar Cabral[1]

Carbon emissions reached a record high in 2024.

In purist terms, we are yet to see a real energy transition. Fossil fuels were supposed to be replaced with renewable energy; yet so far renewable energy hasn't replaced fossil fuels, but rather added to our total energy supply. The energy transition amounts to more of everything.[2]

We are, distressingly, only in the early stages of a transition that should have started decades ago. While this means we will likely pass 2°C of global warming towards the end of the 2030s, it also means that we need to be sober in our assessment of the transition's progress. The early stages of their transition plans always included an increase in energy demand and the probable continuation of fossil-fuel production. They always involved burning oil to build out the transition economy. The transition might only continue to add capacity without retiring fossil fuels; or it may bring about the widely forecast peak of fossil-fuel demand and decline in global emissions. It is too early to tell.

The transition isn't itself energy, however. A transition simply names the process of changing from one state, condition or phase to another. The transition economy isn't solar panels and electric cars: it is an emerging political and economic regime, a project being driven forwards as part of an attempt to resolve several interrelated crises. Its chances of success or failure remain unclear. As a project without clear principles, built with borrowed parts and around conflicting aims, it is a profoundly unsettled political space.

But this creates an opening. The hegemony that broke down after the 2008 financial crisis has yet to be restored. New social and political norms and conventions are not fully established. There is no dominate fraction of Capital; if anything, our ruling classes are increasingly in open conflict. In place of a stable global hegemon, we have rival geopolitical blocs and neo-mercantile conflicts.

As such, rules and norms are all in contestation, while social consent is still under negotiation. Tensions run high, as the political import of each act is laid bare. The process is chaotic, rife with tipping points and cascading impacts, episodes of excess and rupture. At any moment crisis can erupt, dramatically changing the transition's trajectory.

Transitions are no times for compromise. The least bad option is too terrible to contemplate. Compromise means staying within the terms they have set for the transition.

Should the green transition go unchallenged, the future will lock in a green upgrade of existing social, economic and material infrastructures, and consequently one of zero-sum class conflicts in a world marked by persistent undersupply, disruption and a squeeze on living standards.

It will intensify the drive towards green trade wars, monopoly conditions and precarity embedded within the plan, as well as tendencies towards an authoritarian capitalism and a securitised future. All of this in a catastrophic 3°C world.

Some will say we need their investment to enable the innovations required to stop climate change. But this is not true. We already have all the tools we need to not only transform our economies and stop climate change, but to build a better life through a transition that meets our priorities.

If organised, we can disrupt their plans and make this a better world, creating a future where the transition leaves us without one.

But Who?

Books like this one often avoid the question that most of us want answered: who is going to fight the war against green capitalism, and who is going to make the transition we need happen? Who is the 'we' who will make a better world?

The truth is that, like the green transition itself, opposition to it is a work in progress. What we have is an array of movements and a growing sense of running out of time. Our side has yet to be assembled – and we all know it.

Setting out the problem is not enough. There is no single shared interest that will unite us. Even if working-class identity and collective power were somehow formerly produced through the common alignment of simple economic interest, that is no longer the case. And the fracturing of economic interests will only deepen through the green transition. But our unity and power has never been built on purely economic grounds. The hope that capitalism would do the work of organising our class for us has always been futile.

The temptation in books like this one is to set out policies or strategies for a movement still to come. To sidestep the question of how to create a movement in favour of strategic declarations of what to do once we have acquired a measure of counterpower. Or, possibly in worse faith, to lean in on something other than collective power – perhaps the state, should we ever control it, or existing institutions such as unions, or even game-changing new technologies. It should be clear by now that the state is no magic solution, that technology won't save us, and that unions have always been deeply flawed institutions, as often betraying the working class as acting in its interests. None are solutions – all three are terrains of struggle. They can't act for us – we have to act on and through them.

There is a contrary temptation to throw our lot in with the nonhuman forces propelling us towards disaster, and to conceive of climate catastrophe as an agent of social change. To embrace the idea of civilisational collapse as an apocalyptic reckoning, clearing away fossil capitalism and forcing a renewal of humanity.

But the collapse of civilisation will be disappointing.

There will be no sudden cataclysm, no high-octane monster trucks carving their way through sand dunes in a Mad Max dystopia. Collapse isn't an event, but a long-drawn-out process during which the capacity of elites and governments to project power erodes, while trade links and socio-economic complexity unravel. In the past, such disintegrations of regional powers have always been incomplete. And while past collapses did frequently lead to a better life for the bulk of the population, freed from the yoke of local elites, it is unlikely to be liberating for many of us in the over-industrialised Global North.

Of course, the collapse of the ability of the US and Europe to project their power at a global level, to continually renew imperial relations and colonial extraction, would be welcomed by the vast majority of the world's people. Yet here, too, caution is called for. Collapse is messy, and few empires go quietly into decline. All in all, we should not pin our hopes on collapse; it too is an excuse to avoid the hard work of building our own power. It won't save us either.

The subject of the war of green transition is missing. The question is: how to assemble it?

The argument of this book is that we must start to assemble an oppositional force piece by piece. Instead of counting on a single solution or organisation, we need to organise on three terrains and through three distinct (if overlapping) constituencies.

We need to build the environmental movement into a force powerful enough to block fossil capital at its most vulnerable points. Being reasonable, reciting the now obvious facts about catastrophe, pursuing policy – these are dead ends. The blockade is a means of both grinding fossil capitalism down and building a mass movement of ecological militants.

We need to refuse to have the costs of disruption and the transition pushed onto us. The mob and the crowd have been potent historical forces for creating social entitlements. Austerity has devastated the social wage, and the transition threatens to impoverish us even more. Denaturalising the market is a crucial step in challenging the green transition; pushing for price controls from below, to our benefit, builds a new common sense by politicising the economy. It also normalises collective interventions into the market, weaving strong community bonds and a shared sense of refusal.

The transition economy destroys our future. It, as much as climate change, undermines any hope that we might have. Between betrayed promises and the squeeze on our lives, less seems possible each year. We can't begin to organise against their transition plans without creating more space for the future. Without hope, without some chance of something better, all struggles become bitterly defensive. Therefore we anchor our opposition in our plans for a different transition; we begin to make these concrete and real where we live and work. Our transition does not end with our transformed communities; our transformative power begins there. Through community transitions we are able to forge a grounded opposition, one that provides a base from which to organise and operate, as well as consolidate our gains.

The truth is that there is no single figure, no one class fraction or vanguard. We have to do the hard work of cohering an opposition from fragments, across social and political divides, in an increasingly incoherent social field. And we have to do it against the backdrop of incipient fascism and social conflict.

Tell No Lies

The green transition won't stop climate change at safe levels, if such a thing exists. We are likely to land somewhere between 2°C and 3°C – catastrophic levels by any measure.

The impacts will be terrible. Already at somewhere around 1.5°C the world is suffering an adaptability crisis, with some areas becoming barely habitable. At 3°C, deserts and arid zones will spread, rainfall will become unreliable and scarce in many places (or torrential in others); heat will make whole regions brutally hostile to life during the day. All of this will put massive strains on existing infrastructure and services. Breakdown will become an everyday experience, rendering more places barely liveable.

There will be a constant battle to force governments to act, amid efforts to prevent the various corporations and asset management funds that control essential infrastructure from ditching us where it's no longer profitable to provide services. Faced with the daunting scale of not only intervening into the transition economy to reshape it to our plans and needs, but of also fighting a rearguard action against our abandonment by both state and big business, it could be tempting to turn inwards. To look to our own survival. To build the resilience and autonomy of our immediate communities.

But our mere survival constitutes no threat. And without opposition, only the worst possible transition will take place. Should we resign ourselves to an abandoned future, fitfully working to install increasingly faulty solar arrays, it is highly likely that political mobilisation will become the preserve of an ever-more reactionary far right.

While we know that government isn't coming to save us, we also need to understand that no disaster will come to save us. There is no hope to be found in fantasies of collapse. There are no easy victories to be claimed. Instead we must focus on the hard work of organising, and take courage from each other. We must grasp hold of the transition, to reshape it to our ends. We must become the climate disruption: the disaster to their plans.

Notes

Introduction: The Debate Is Over – on to the Green Transition

1   The term Global North is not without controversies and complexities, but for the purposes of this book it is sufficient and has the benefit of foregrounding the ongoing imperial and neo-colonial relations of the transition economy. However, this doesn't mean that everyone living in the Global North is wealthy, nor that wealth and the imperial exercise of power are absent from the Global South.

2   U. Brand and M. Wissen, *The Imperial Mode of Living: Everyday Life and the Ecological Crisis of Capitalism*, Verso Books, 2021.

3   DBT, Wales Office, 'Tata Steel / Port Talbot steelworks Q&A', 19 February 2024.

4   A just phase-out of fossil fuels, one that enables economic growth and development in the Global South, requires the world's wealthiest countries to end their fossil fuel production by the early 2030s.

5   L. Dai, L. Clements, A. Meng, B. Schuck and J. Kooroshy, 'Investing in the green economy 2024: Growing in a fractured landscape', London Stock Exchange Group plc, 2024.

6   S. Mundy, 'Transition to clean energy falters as green tech funding falls short', *Financial Times*, 11 November 2024.

1. The Squeeze

1　Press Team, 'Thirty-six Percent of US Families Skipped Meals for Financial Reasons, Dunnhumby Study Finds', Dunnhumby, 2 August 2023.

2　L. Gonçalves and M. Long, 'US food insecurity rate rose to 13.5% in 2023 as government benefits declined and food prices soared', The Conversation, 5 September 2024.

3　S. O'Connor and J. Burn-Murdoch, 'Left behind: can anyone save the towns the UK economy forgot?', *Financial Times*, 16 November 2017.

4　G. Wearden, 'Britons "need to accept" they're poorer, says Bank of England economist', *Guardian*, 25 April 2023.

5　R. Patel and J. W. Moore, *A History of the World in Seven Cheap Things: A Guide to Capitalism, Nature, and the Future of the Planet*, University of California Press, 2018.

6　A. Ortiz-Bobea, T. R. Ault, C. M. Carrillo, R. G. Chambers and D. B. Lobell, 'Anthropogenic climate change has slowed global agricultural productivity growth', *Nature Climate Change*, 11(4), 2021, pp. 306–12.

7　T. Kompas, T. N. Che and R. Q. Grafton, 'Global impacts of heat and water stress on food production and severe food insecurity', Scientific reports, 14(1), 2024.

8　Ibid.

9　T. Frank, 'Climate Change Is Destabilizing Insurance Industry', *Scientific American*, 23 March 2023.

10　T. Seal, 'Insurance czar has "harsh" message about climate: You may just have to move', Bloomberg, 23 September 2024.

11　C. Hickman, E. Marks, P. Pihkala, S. Clayton, R. E. Lewandowski, E. E. Mayall, B. Wray, C. Mellor, and L. van Susteren, 'Climate anxiety in children and young people and their beliefs about government responses to climate change: a global survey', *Lancet Planetary Health*, 5(12), 2021.

12 C.A. Ogunbode, R. Doran, D. Hanss, M. Ojala, K. Salmela-Aro, K.L. van den Broek, N. Bhullar, S.D. Aquino, T. Marot, J.A. Schermer, A. Wlodarczyk, S. Lu, F. Jiang, D. A. Maran, R. Yadav, R. Ardi, R. Chegeni, E. Ghanbarian and R. Najafi, 'Climate anxiety, wellbeing and pro-environmental action: correlates of negative emotional responses to climate change in 32 countries', *Journal of Environmental Psychology*, 84, 2022, p. 101887.

13 L. Berlant, *Cruel Optimism*, Duke University Press, 2020.

14 A. Malm, *Fossil Capital: The Rise of Steam Power and the Roots of Global Warming*, Verso Books, 2016.

2. Working the Transition

1 R. Baldwin, 'China is the world's sole manufacturing super-power: a line sketch of the rise', CEPR, 17 January 2024.

2 B. Moshinsky, 'This one statistic sums up why UK steel can't compete with China', *Business Insider*, 31 March 2016.

3 A. Lawson, 'World's largest solar manufacturer to cut one-third of workforce', *Guardian*, 18 March 2024.

4 D. Fickling and T. Culpan, 'Solar success is a curse for China's manufacturers', Bloomberg, 11 March 2024.

5 'Nearly all US farms are family farms, USDA says', Farm Policy News, 13 December 2024.

6 British landfill sites are responsible for double the carbon emissions of Tata Steel's Port Talbot site.

7 AFP, 'China's share of global manufacturing jobs to rise by 2050, study finds, even as US, EU seek less reliance on its products', Hong Kong Free Press, 30 October 2023.

8 J. Ambrose, 'China to head green energy boom with 60% of new projects in next six years', *Guardian*, 9 October 2024.

9 L. Harris, ' "It's boom time": Renewable growth is faster in the global south than in rich countries', *Financial Times*, 16 October 2024.

10 K. Stancil, 'Teachers union in Ohio went on strike for students –
    and won', Common Dreams, 29 August 2022.

11 B. Johnson, 'NEU: Campaign for legal maximum working
    temperature', Twinkl News, 17 April 2024.

12 J. Knutson, 'Over 70% of world's workforce exposed to excessive
    heat each year, UN finds', Axios, 22 April 2024.

13 A. Baker, 'Extreme heat is endangering America's workers – and
    its economy', *Time*, 3 August 2023.

14 'Heat deaths at work up by 40% in the EU', European Trade
    Union Confederation, 26 April 2024.

15 EJOLT, *'Mining conflicts in Latin America'*, Environmental
    Justice Atlas, accessed 20 August 2024.

16 M. Arboleda, *Planetary Mine: Territories of Extraction under
    Late Capitalism*, Verso Books, 2020.

17 'Private security services market size, share, industry analysis,
    trends, growths, forecasts, 2032, Zion Market Research, 2024.

18 W. I. Robinson, *The Global Police State*, Pluto Press, 2020.

19 D. Graeber, *Bullshit Jobs: A Theory*, Penguin Books, 2018.

20 'Global Manufacturing Has Likely Peaked, Even in Poor Countries,
    New Study Finds', Center For Global Development, 30 October 2023.

3. The Transition Economy

1 Of course, it's entirely possible that, in response to social and
   political conflicts thrown up by both climate change and the
   transition, governments abandon transition economics, just as it's
   possible that changes to carbon sinks and the crossing of tipping
   points push us well past 3°C.

2 While firmly anti-neoliberal in character, there was more than a
   little similarity between the framing of these policies as a war-
   footing effort to tackle climate change and the very neoliberal
   emphasis on government funding for military programmes and
   military accumulation.

3 Obviously, carbon credits and carbon markets foreshadow these developments, but they form only part of a much broader process of privatisation.

4 R. Brenner, *The Economics of Global Turbulence: The Advanced Capitalist Economies from Long Boom to Long Downturn, 1945–2005*, Verso Books, 2006.

5 G. Arrighi, *Adam Smith in Beijing: Lineages of the Twenty-First Century*, Verso Books, 2009, p. 132.

6 R. Solow, 'We'd better watch out', *New York Times Book Review*, 12 July 1987, p. 36.

7 'Incremental innovation' constitutes around 75 per cent of all innovation work.

8 C. Doctorow, '"Enshittification" is coming for absolutely everything', *Financial Times*, 7 February 2024.

9 A. Benanav, *Automation and the Future of Work*, Verso Books, 2020.

10 D. Vollrath, *Fully Grown: Why a Stagnant Economy Is a Sign of Success*, University of Chicago Press, 2019.

11 N. Fraser, *Cannibal Capitalism: How Our System Is Devouring Democracy, Care, and the Planet and What We Can Do About It*, Verso Books, 2023.

12 J. Crary, *24/7: Late Capitalism and the Ends of Sleep*, Verso Books, 2013.

13 R. Patel and J. W. Moore, *A History of the World in Seven Cheap Things: A Guide to Capitalism, Nature, and the Future of the Planet*, University of California Press, 2017.

14 B. Christophers, *The Price Is Wrong: Why Capitalism Won't Save the Planet*, Verso Books, 2024.

15 D. Gabor, 'The Wall Street Consensus', *Development and Change*, 52(3), 2021, pp. 429–59.

16 B. Christophers, *Rentier Capitalism: Who Owns the Economy, and Who Pays for It?*, Verso Books, 2020.

17   'The net-zero transition: what it would cost, what it could bring', McKinsey & Company, January 2022.

18   For a useful corrective to many of the more hyperbolic accounts, see A. Tooze, 'Whose century?', *London Review of Books*, Vol. 42, No. 15, 2020.

19   A. Tooze, 'Great Power Politics', *London Review of Books*, Vol. 46, No. 21, 2024.

20   A. He, 'In the Global AI Chips Race, China Is Playing Catch-Up', Centre for International Governance Innovation, 18 September 2024.

21   T. Sahay, 'A New Non-Alignment', Phenomenal World, 9 November 2022.

22   J. Hickel, D. Sullivan and H. Zoomkawala, 'Plunder in the Post-Colonial Era: Quantifying Drain from the Global South Through Unequal Exchange, 1960–2018', *New Political Economy*, 26(1), 2021, pp. 1–18.

23   'Data show Global South is in worst debt crisis ever, with another lost decade looming', Bretton Woods Project, 13 December 2023.

24   Urbanisation and industrialisation are linked phenomena, with urbanisation often being largely a consequence of industrialisation and government policies designed to drive industrial-led economic growth, coupled with the enclosures necessary to ensure a steady supply of raw materials.

25   D. Green, 'Are food prices becoming more volatile? Yes, says the FAO (but it doesn't know what to do about it)', Oxfam, 11 February 2011.

26   J. O'Connor, 'On the two contradictions of capitalism', *Capitalism Nature Socialism*, 2(3), 1991, pp. 107–9.

27   Patel and Moore, *History of the World in Seven Cheap Things*.

28   I. M. Weber and E. Wasner, 'Sellers' inflation, profits and

conflict: why can large firms hike prices in an emergency?',
Review of Keynesian Economics, 2023.

4. Blockading the Transition

1 Earnings from fossil fuels for governments, from direct owner-ship to taxation to licencing fees, amounts to around US\$1.5–2 trillion per year.

2 This fact – that we must stop burning fossil fuels to stop climate change – is now so established that we can dispense with justifi-cations for it.

3 The figures for such reductions are annual global cuts of between 7 and 10 per cent per year, far above anything found outside of profound economic contractions such as depressions and economic collapse. And this is for an unjust transition, that abandons any equitable sharing of the remaining 'carbon space' with poorer nations who are the least responsible for climate change. The equitable sharing of our remaining climate space would require countries like the USA and Britain to cut their emissions, starting immediately, by 10 to 15 per cent per year.

4 IEA, *World Energy Employment 2022*, IEA, September 2022.

5 B. Morton, 'New oil and gas ban threatens jobs, unions warn', BBC News, 9 September 2024.

6 J. Meadway, 'Is a "green state" the answer to the climate crisis?', *New Statesman*, 10 November 2021.

7 A. Malm, *How to Blow Up a Pipeline*, Verso Books, 2021.

8 Ibid., p. 154.

9 L. Temper, S. Avila, D. D. Bene, J. Gobby, N. Kosoy, P. L. Billon, J. Martinez-Alier, P. Perkins, B. Roy, A. Scheidel and M. Walter, 'Movements shaping climate futures: A systematic mapping of protests against fossil fuel and low-carbon energy projects', *Environmental Research Letters*, 15(12), 2020, p. 123004.

10 M. Lynas, *Our Final Warning: Six Degrees of Climate Emergency*, HarperCollins, 2020.

11 J. Clover, *Riot. Strike. Riot: The New Era of Uprisings*, Verso Books, 2019.

12 T. Mitchell, *Carbon Democracy: Political Power in the Age of Oil*, Verso Books, 2011.

13 Total gas production is also geographically concentrated, with over two thirds of production located in just five states, with 20 per cent coming from the Permian Basin shale field alone. This concentration is also reflected in the industry structure, with the top ten companies accounting for 31 per cent of total US natural gas production in 2009.

14 K. Rives, 'US has 133 new gas-fired plants in the works, putting climate goals at risk', SP Global, 15 May 2024.

15 There is not the space here to debunk claims that aviation can be made sustainable, specifically through ammonia or 'sustainable aviation fuels'. The latter, which some governments such as the British are subsidising as frontier technologies, actually make emissions worse.

5. Refusing the Price of Their Transition

1 G. Petro, 'Electric Vehicle Boom Hits a Wall of Consumer Ambivalence', Forbes, 7 March 2024.

2 The difference between the two is fuzzy, but essentially consumer choice is about what we buy, while behavioural change is about the ways in which we use the things we buy. A consumer chooses an electric car, but driving less frequently is a behavioural change. Switching to a heat pump is a consumer choice, but lowering the thermostat is a behavioural change.

3 J. Lovelock, *The Revenge of Gaia: Why the Earth Is Fighting Back, and How We Can Still Save Humanity*, Penguin Books, 2007; G. Monbiot, *Feral: Rewilding the Land, the Sea, and Human Life*,

University of Chicago Press, 2014; R. Scranton, *Learning to Die in the Anthropocene: Reflections on the End of a Civilization*, City Lights Books, 2015.

4 C. Lodziak, *The Myth of Consumerism*, Pluto Press, 2002.

5 Brand and Wissen, *The Imperial Mode of Living*.

6 C. Dewar, 'Climate change protesters "steal food from Tesco" to give to public', STV News, 21 February 2024.

7 Office for National Statistics, crime statistics presentation and communication review, current and upcoming work: October 2024.

8 BRC, *Crime Survey 2024 Report*, 14 February 2024.

9 Capital One Shopping, 'Retail Theft (Shoplifting) Statistics', 26 September 2023.

10 C. Blanco, J. Grant, N. M. Petry, H. B. Simpson, A. Alegria, S. M. Liu and D. Hasin, 'Prevalence and correlates of shoplifting in the United States: results from the National Epidemiologic Survey on Alcohol and Related Conditions (NESARC)', *American Journal of Psychiatry*, 165(7), July 2008, pp. 905–13; more recent surveys have found a similar proportion of the population shoplifting. See Capital One Shopping, 'Retail Theft (Shoplifting) Statistics'; C. Wolfe, 'New Survey Shows More Than 1 in 5 Americans Have Shoplifted', Loss Prevention Media, 7 August 2024.

11 E. P. Thompson, 'The moral economy of the English crowd in the eighteenth century', *Past and Present*, 50(1), February 1971, pp. 76–136; D. Harvie and K. Milburn, 'The moral economy of the English crowd in the twenty-first century', *South Atlantic Quarterly*, 112(3), 2013, pp. 559–67.

12 S. Wheeler and J. Squire, 'Food Price Hikes, Social Composition and Auto-Reduction', Notes From Below, 21 September 2023.

13 Don't Pay (n.d.), *Don't Pay*. Available at: dontpay.uk.

14 The death of the British queen also played a role in interrupting the campaign's momentum.

15 M. Simpson, 'Britain's Historic Wave of Student Rent Strikes', *Tribune*, 21 January 2021.

16 B. Trott, 'Walking in the right direction?', turbulence.org.uk, 2021, accessed 13 November 2024.

6. Community Transitions and Counter-planning

1 J. F. Mcalevey, *No Shortcuts: Organizing for Power in the New Gilded Age*, Oxford University Press, 2016.

2 Ibid., p. 61.

3 F. Lister-Fell, 'When bosses tried to sack them, these automotive workers took over their factory', Progressive International, 19 July 2023.

4 After a series of sales and mergers starting in 1996, Lucas Aerospace is now owned by Collins Aerospace, a subsidiary of RTX Corporation, and is one of the largest suppliers of defence and aerospace technology in the world.

5 A. Smith, 'Technology networks for socially useful production', *Journal of Peer Production*, Issue 5, 2014.

6 S. Federici, *Caliban and the Witch*, Autonomedia, 2004; T. Bhattacharya (ed.), *Social Reproduction Theory: Remapping Class, Recentering Oppression*, Pluto Press, 2017.

7 M. Mies, *Patriarchy and Accumulation on a World Scale: Women in the International Division of Labour*, Bloomsbury Publishing, 2014; Patel and Moore, *A History of the World in Seven Cheap Things*.

8 N. Cox and S. Federici, *Counter-Planning from the Kitchen: Wages for Housework, a Perspective on Capital and the Left*, Falling Wall Press, 1975.

9 It's more than ironic that as I write these words Amazon have just announced that they will be building new cloud and data processing centres in Britain, creating 14,000 new jobs, no doubt with the same wages and conditions as their other warehouse jobs.

This week it is also expected that Tata Steel will formally announce over 2,500 job losses in Port Talbot.

10 M. Burgmann and V. Burgmann, 'Green Bans movement', dictionaryofsydney.org, 2011, accessed 20 December 2024.

11 M. Burgmann and V. Burgmann, *Green Bans, Red Union: Environmental Activism and the New South Wales Builders Labourers' Federation*, UNSW Press, 1998.

7. Struggling for a Liveable World

1 'Hide nothing from the masses of our people. Tell no lies. Expose lies whenever they are told. Mask no difficulties, mistakes, failures. Claim no easy victories . . .' from A. Cabral, *Revolution in Guinea: An African People's Struggle. Selected Texts*, Stage 1, 1971, pp. 70–2.

2 J. B. Fressoz, *More and More and More: An All-Consuming History of Energy*, Allen Lane, 2024.